RFID
for Energy & Utility Industries

RFID
for Energy & Utility Industries

Dipankar Sen • Prosenjit Sen • Anand M. Das

PennWell®

PennWell Corporation
1421 South Sheridan Road
Tulsa, Oklahoma 74112-6600 USA

800.752.9764
+1.918.831.9421
sales@pennwell.com
www.pennwellbooks.com
www.pennwell.com

Marketing Manager: Julie Simmons
National Account Executive: Barbara McGee

Director: Mary McGee
Managing Editor: Steve Hill
Production Manager: Sheila Brock
Production Editor: Tony Quinn
Book Designer: Susan Ormston
Cover Designer: Alan McCuller

Library of Congress Cataloging-in-Publication Data

Sen, Dipankar.
RFID for energy & utility industries / Dipankar Sen, Prosenjit Sen, Anand M. Das.
p. cm.
Includes bibliographical references and index.
ISBN 978-1-59370-105-5
1. Radio frequency identification systems. 2. Business logistics. 3. Electric utilities--Materials management. 4. Water utilities--Materials management. 5. Gas companies--Materials management. 6. Inventory control--Technological innovations. I. Sen, Prosenjit. II. Das, Anand M. III. Title. IV. Title: Radio frequency identification for energy and utility industries.
TS160.S45 2008
628.1068'7--dc22 2008033719

Printed in the United States of America

1 2 3 4 5 13 12 11 10 09

DEDICATION

We are particularly indebted to Amelie and Rini; to Nisha, Diya and Durba; and Priyadarshini, Shivani and Annapurna for their support during this effort.

Dipankar Sen
Prosenjit Sen
Anand M. Das

CONTENTS

ILLUSTRATIONS

Figures

Tables

INTRODUCTION

RFID Application to Improve Asset Management at Water, Electric, and Gas Utilities

RFID and wireless technologies are gradually being introduced in the utility industry. Innovations made in the defense and logistics industries can now be adapted for the nature of operations at water, electric, and gas utilities. In this book, the authors apply the knowledge of applications in various industries to identify problems that can be solved using RFID and wireless technologies.

One challenge in adapting the RFID technologies to utilities is to find the right mix of technologists and utility managers who can understand the technology and adapt it to the problems faced at utilities. For this, we have found it necessary to introduce the technology and the assets within the utility at a level that can be understood by technologists, plant engineers, or managers. In chapters 1 through 4, we introduce the reader to the technology. This is followed by a discussion of applications in three types of utilities. The discussion of applications is kept at a level such that a manager at a water utility will also find the discussion regarding electric and gas utilities interesting, and vice versa. Hopefully, this will help the manager and the technologist understand how applications developed in one sector of the utility industry can be applied to others. To this effect, we have also included a glossary of acronyms in the appendix to help the reader with the terminologies.

Chapter 1 introduces the RFID basics of RFID technology. It covers the types of tags available and the components that make up the RFID system. These include the passive and active tags, and the forms, shapes, and ruggedness of tags. The chapter includes software applications necessary to support the RFID network.

Chapter 2 introduces the wireless infrastructure that is necessary to transmit the information from the RFID tags to the receiving devices. It looks at four categories of wireless networks with increasing geographic reach, beginning with personal area networks and increasing in size to local, metro, and wide area networks. We discuss technologies such as Bluetooth, Infrared, ZigBee, and Wi-Fi, together with a discussion of the relevant standards of communication.

Chapter 3 gets into the development of applications for mobile asset management. We look at applications such as materials management, especially where materials have a finite shelf life, hazardous waste management and shipping, tools and parts management, and vehicle and fleet management. Most utilities have an immediate need and can garner benefits from application of mobile asset management in some parts of their organization.

Chapter 4 looks into how RFID labeling can be applied for enhancing the information available in the field from the tagging of fixed assets. Also, it looks into how it can be used to enhance the process of locating hidden assets.

In chapter 5, the authors move from the technology to application within the water utility. The types of assets within the water and wastewater industries are described. The intention is to familiarize the reader with the assets. Based on this, the applications of RFID for fixed asset management, mobile asset tracking, vehicle and fleet management, and infrastructure security and monitoring applications such as automated meter reading are described.

In chapter 6, the discussion focuses on the electric utilities. The assets in all three functional areas—generation, transmission, and distribution—are described. The intent is not to be an encyclopedia of assets but to focus on some of the key assets in each functional area. The various types of problems where RFID applications may be applied or are already being applied are introduced next. This is followed by a discussion of solutions.

Chapter 7 follows the same format for the natural gas utilities and the natural gas industry. The description of assets begins with extraction and processing and then moves to transmission and distribution. Various problems are introduced next followed by a discussion of solutions around improvements that can be made to fixed and mobile asset management, access control and security, and monitoring.

Appendix A includes a summary on how to compute costs and benefits in the evaluation of an RFID project. It has been structured in a manner to help increase the return on investment in RFID- and wireless-based technologies.

Appendix B includes a glossary of terms used throughout the text.

1 RFID TECHNOLOGY

Introduction

RFID or *radio frequency identification* refers to the method of identification of an object by attaching a tag to it that uses radio frequency to communicate. The RFID tag has a unique identifier to represent and identify the object it is attached to; certain RFID tags can store additional data in memory.

RFID belongs to a class of technologies called *AIDC* or *automatic identification and data capture.* AIDC technologies have been in existence since the 1930s and include technologies such as optical character recognition (OCR), bar code, magnetic ink character recognition, electronic article surveillance, and biometric identification. Bar code has been so far the most pervasive AIDC technology; however, it is widely expected that RFID will one day take its place.

Even though RFID technology has been around for many years, until a few years ago it was mainly used by companies for proprietary purposes. In the 1960s, the U.S. government started using RFID to track and monitor nuclear and certain other hazardous materials. In recent years, the adoption of RFID technology by Wal-Mart and the Department of Defense (DOD) for supply chain management has resulted in increased interest among the business community. It is now expected that in a few years the prices of tags will drop to a point where it will become feasible to replace bar codes with this technology and allow item-level tracking and identification of items.

Wal-Mart has publicly announced that it requires its suppliers to tag cases and pallets with passive RFID tags. In commodity-focused links in the supply chain, the DOD is aggressively pursuing the adoption of passive RFID technologies. As set forth in a policy update on July 30, 2004, Michael Wynne, Acting Under Secretary of Defense for Acquisition, Technology, and Logistics [USD (AT&L)], has affirmed that DOD suppliers will ship orders to DOD customers with passive RFID tags affixed.

An RFID system is essentially comprised of three types of components—RFID tags, RFID readers, and the host computer. An RFID tag is a transponder that can be attached to or implanted into products, animals, or other objects so they can be uniquely identified. Tags are read using an RFID reader, a device that uses radio frequency signals to recognize the presence of tags when they are within its read range. The RFID reader is also capable of writing into an RFID tag if the RFID tag is writeable.

When an RFID tag is detected by an RFID reader, this is called an *observation.* When there is a change in the state of an observation or when an observation differs from its previous observation, it is called an *event.* The RFID readers pass on the observations or events for further processing to the host computer. The host computer is hardware that runs business applications to take this RFID data and process it along with other business data to generate reports. Such reports provide visibility into business processes. Figure 1–1 is a diagram of an RFID system.

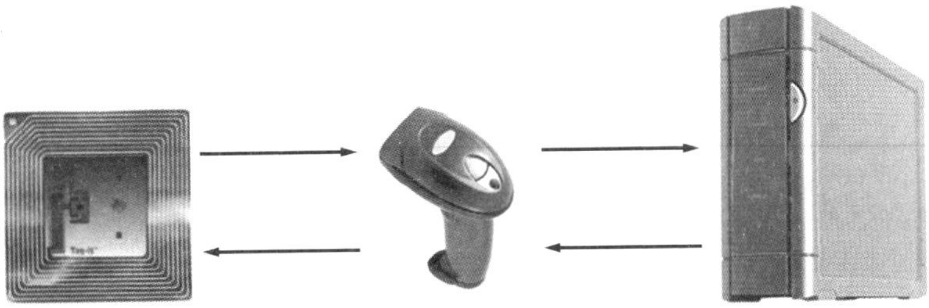

Fig. 1–1. The RFID system

One may wonder, since there are several types of AIDC technologies already available and bar codes are so popular among businesses, why there is a need for RFID-based identification. Is it feasible to expect that RFID may replace or enhance information available from bar codes? After all, the bar code technology is mature, inexpensive, and extensively used by businesses.

The current applications of RFID show that RFID-based identifiers can be used to provide a significantly superior solution compared to bar codes. A bar code can be read only when it is directly visible to the bar code reader. In other words, a *line of sight* is required between the bar code and the reader. If the bar code is concealed or placed inside the item, it cannot be read.

An RFID tag, on the other hand, has no such limitation. An RFID tag placed inside an item can be read perfectly fine as long as the tag is within the read range of the antenna of the RFID reader. This allows a higher level of automation while implementing solutions for tracking items. Some of the other advantages of using RFID-based identification over bar codes are as follows:

- **Item-level tracking.** Since an RFID tag can have 96-bit identifiers or 128-bit identifiers (as per GEN2 standards), it allows tracking at the item level instead of the item-type level. For example, if one applies RFID tags on 1,000 ink jet printers having model number HP Business Inkjet 1100, it is possible to uniquely identify each of these units individually instead of having to label them all with the same identifier to represent that they are all HP Business Inkjet 1100 printers.

- **Reads multiple tags.** An RFID reader can read multiple RFID tags simultaneously, therefore making it possible to implement a more automated solution that reads faster than bar code–based solutions.
- **Stores additional information.** In the case of bar codes, once the tag is printed it cannot be changed anymore. RFID tags can be of different types—some are *read only*, while others have additional memory that can be written into over and over again. Such *writeable* tags allow additional business data to be stored. For example, if a writeable RFID tag is attached to a tool, every time the tool is checked out or checked in by a technician, an RFID reader can write the name of the technician and the time checked in or out.

Applications of RFID

As the cost of RFID tags and RFID readers goes down over time, businesses increasingly will use this technology for business process efficiency. Here are a few possible applications:

- **Manufacturers and retailers.** Track items in real time as they move through the supply chain. This would enable improved planning, forecasting, and management of supplies.

- **Water or gas utilities.** Track tools as they are checked in and out by the maintenance personnel, preventing delay due to unavailability of the right tools and resulting in better utilization of tools.
- **Power plants.** RFID-based smart cards for access control of authorized personnel.
- **City or county.** Tag hazardous material to ensure proper management and disposal by businesses as per the regulations of the city or county.
- **Industrial facility.** Tag fixed assets such as valves and switches for efficient maintenance and operations.
- **Data centers.** Tag tapes and other backup devices to prevent misplacements and also to ensure that they are accessible only to authorized personnel. Similarly, laptops, disks, and other data storage devices can also be tagged.
- **Department of Defense.** The Defense Logistics Agency has committed to read and apply Electronic Product Code (EPC)–compliant passive RFID tags on all inbound and outbound shipments.

Architecture of an RFID System

According to ANSI/IEEE 1471:: ISO/IEC 40210, the architecture of a system is defined as "the fundamental organization of a system, embodied in its components, their relationships to each other and the environment, and the principles governing its design and evolution."[1]

The architecture of an RFID system can vary considerably depending on its application. The number of readers, the number of tags read by a reader at any given moment, the location and accessibility of the readers and other hardware, and the frequency of the events generated are all criteria that go into designing the architecture of the RFID system. Some systems are set up for centralized processing of data, while others are more distributed.

The RFID system can be proprietary or standards based; it can be implemented as a solution within an enterprise or as one that allows sharing of information across enterprises. It can be used for locating fixed assets, tracking mobile assets, access control, and more.

The following are some of the capabilities that one would expect in an RFID system:

- **Encoding and attaching of RFID tags.** Encoding involves generating a unique identification number (using numbers and letters) and then writing it into the tag such that it can be read by an RFID reader. The process of generating a unique identification number is done using a certain item numbering or identification generation scheme. A sample encoding of 96-bit tag construct is as follows:

Header	Filter	Company Code	Serial Number
8 bits	4 bits	48 bits	36 bits

Fields:

(a) Header—specifies that the tag data is encoded as a 96-bit tag construct, use binary number: 1100 1111

(b) Filter—identifies a pallet, case, or UID (unique identifier) item associated with tag, represented in binary number format using the following values:

- 0000 = pallet.
- 0001 = case.
- 0010 = UID item.
- All other combinations = reserved for future use.

(c) Company code—identifies the vendor and ensures uniqueness of serial number across all vendors represented in ASCII format.
(d) Serial number—uniquely identifies up to 236 = 68,719,476,736 tagged items, represented in binary number format.

Next consider the process of attaching the RFID tag to an object. An RFID printer can be used to first encode the RFID tag, then print on it, and finally apply it to the item. An adhesive is often used to attach the RFID tag to the object. The RFID tag includes a chip and an antenna, which are delicate and can sometimes fail while the tag is printed or attached to an object. Therefore, it is important to have a two-step verification process to ensure that an RFID tag attached to an object is not defective. This involves first testing the tag before writing into it to make sure that it is not defective initially. Following that the RFID printer writes into the tag, prints it, and applies it to the item. Once it is applied to the item, it is tested again to ensure that it did not become defective during the process of printing and application.

- **Reading of RFID tags.** The tags should be readable by RFID readers (and also writeable in the case of writeable RFID tags) as per the requirements of the application.

- **Processing of RFID data by business applications.** The RFID data read by the RFID readers should be channeled to business applications for further processing. The business applications consider the RFID events in the context of other business data to draw meaningful conclusions and generate reports. Existing business applications may need to be modified to accommodate tracking at the item level instead of the item-type level that has been prevalent so far due to bar code–based tagging. Item-level tracking allows every individual item to be uniquely identified.

- **Manageability.** This refers to the ability to manage the hardware and software components of the RFID system electronically in order to allow failure or malfunction of a component to be detected instantly. An automated management scheme should allow for the functionality of a failed component to be taken over by another component of the system until the failed component is repaired or replaced, to minimize downtime.
- **Security.** It is important to ensure that the data stored in the tags, the servers, and any intermediate hardware are secure so the data is not compromised to unauthorized personnel. For this purpose, encryption of the data stored in the tags and other hardware is essential. It is also important to encrypt data as it is exchanged between the various components of the RFID system. Additionally, user authentication functionality is also required so only authorized users are given access to data from the RFID system.

Figure 1–2 shows the components of an RFID system. The bottom two layers, "chips/tags" and "hardware/readers" refer to the RFID hardware. "Data management" and "applications" refer to the software that is used to process the RFID data to provide visibility into a business process. The "applications" layer is referred to as *receiving applications* in the rest of this chapter.

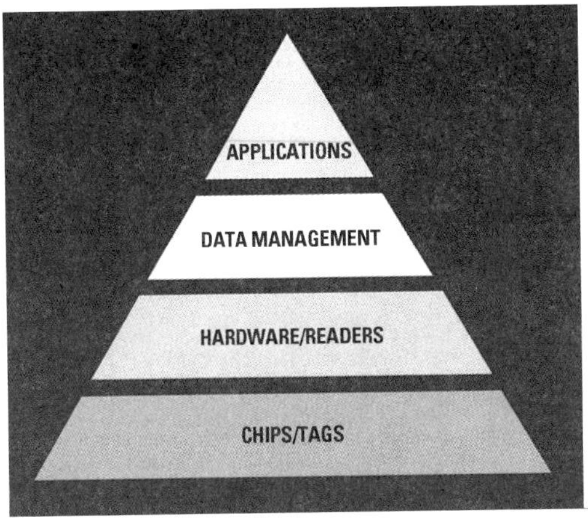

Fig. 1–2. The RFID stack.
Source: Forrester Research

RFID Tags

RFID tags are unique identifiers, and the use of them allows items to be uniquely identified. RFID tags come in different shapes and sizes—paper labels, buttons, inside a car key, and others (figs. 1–3 and 1–4).

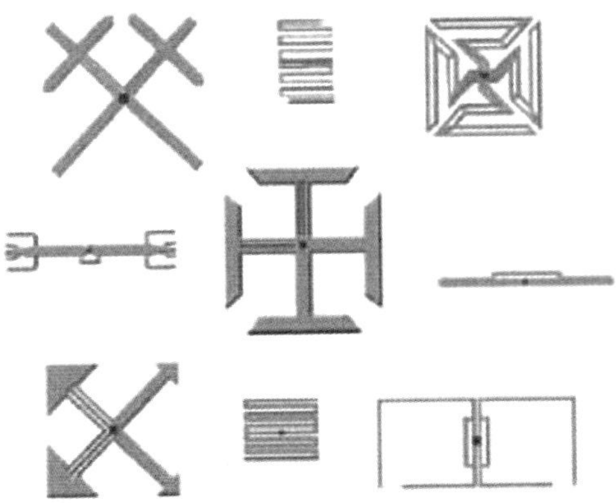

Fig. 1–3. RFID tags from Symbol Technologies

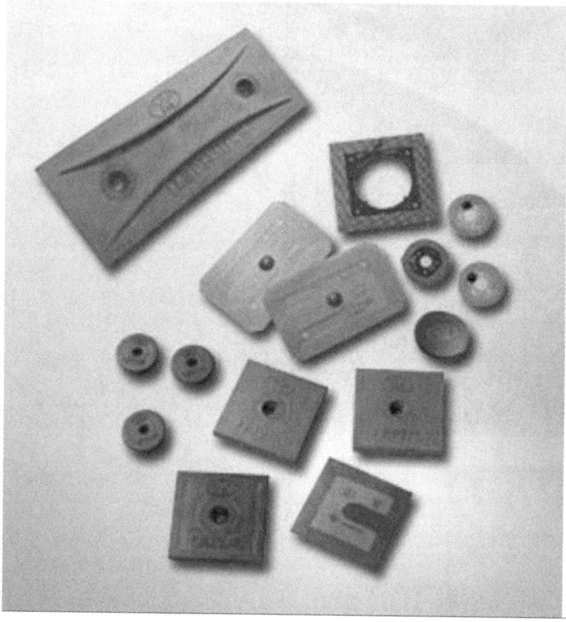

Fig. 1–4.
RFID tags from Escort Memory Systems

Tags can be categorized into different types based on form factor, the frequency they use to communicate with the readers, and their source of power.

A tag is comprised of an antenna, a microchip, and sometimes a battery (fig. 1–5). The antenna is used for communicating with an RFID reader, while the microchip is used to store the identification number (and sometimes other data). Sometimes, a tag can have its own battery to provide power to the microchip for reading and writing (fig 1–6). In addition to these components, tags come with packaging depending on the type of application and the environment where it is used.

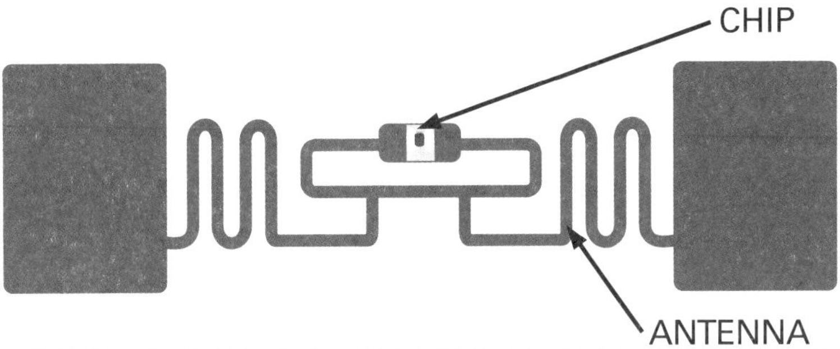

Fig. 1–5. Passive RFID tag ALN-9562 - "Squiggle®-SH." *Source: Alien Technology*

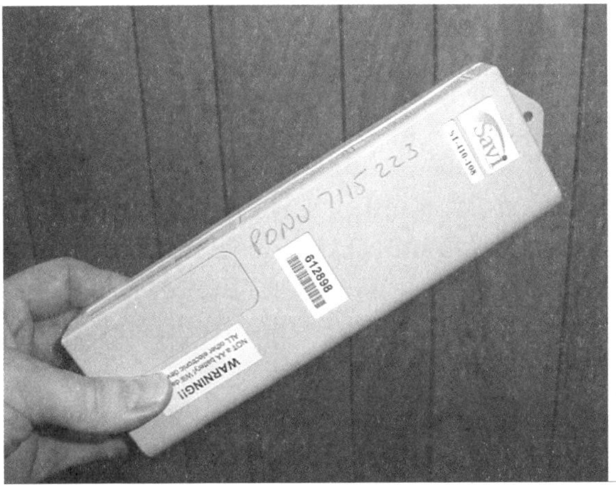

Fig. 1–6. SAVI 410 Active RFID Tag. *Source: SAVI Technology*

The most common RFID tags come as labels that are similar to bar codes. Such tags are usually visible and have an adhesive layer at the back so they can be affixed to an item. Tags can also be embedded in a key (for example, a car key), key chain, watch, a piece of clothing, and more. These are usually concealed and cannot be visually detected. Tags can also be present in buttons, glass capsules, or access cards.

The type of packaging can vary quite a bit depending on the environment where the tags will be used. In environments such as outdoors or inside a chemical factory, tags need to be packaged with adequate protective material to prevent any damage. Figure 1–7 gives an example of a tag that can operate in a high-temperature environment.

Fig. 1–7. EPC UHF 900 MHz micro metal high-temperature tag from Dynasys. *Source: RFIDusa.com*

Active or passive tags

RFID tags can be passive, semi-passive, or active. Passive tags have no internal power source, while semi-passive and active tags have a small battery as power source.

Passive tags are activated by the radio frequency signal from an RFID reader. The electromagnetic field of an RFID reader provides energy to active the tag, so the data (identifier and more) stored in the chip is read and returned to the RFID reader.

An active tag has its own battery power for reading from or writing into the microchip and to communicate the data to the RFID reader. An active tag, since it has its own battery, is more expensive than a passive tag but has a higher read range compared to passive tags. Active tags can have a range of 100 meters (m) and a battery life even as high as 10 years. Active tags can be of two types—one where the tag responds only when it is queried by an RFID reader and the other where the tag itself transmits periodically. The latter is also referred to as an *active beacon tag*.

Active tags usually do not have issues in environments having water and metals, unlike certain passive tags. Active tags also can store more data than passive tags and are often used in conjunction with other systems, devices, and technologies.

Active tags are used along with sensors that measure humidity, temperature, light radiation, and radiation. The following are some of the technologies that are or soon will be used by active RFID, or combined with it for different applications:

- Bluetooth
- Wi-Fi
- WiMax
- GPS
- GSM
- ZigBee
- GPRS
- Infrared

Semi-passive tags have their own battery power but are usually in sleep mode. When an RFID reader tries to read such a tag, the radio frequency signal transmitted by the reader activates the battery that then provides the power to read the microchip. The data transmission is accomplished using power from the RFID reader, similar to that of passive tags. These tags are more expensive than passive tags but are designed to conserve the battery and therefore last longer than active tags. Semi-passive tags are also known as *battery-assisted tags.*

Operating frequency

The operating frequency is the electromagnetic frequency used by the tag to communicate with the RFID reader. The frequency ranges used in the case of RFID are classified as low frequency (LF), high frequency (HF), ultra-high frequency (UHF) and microwave. Figure 1–8 provides details on the frequency ranges versus the maximum read range for such tags.

Frequency	Regulation	Range	Data Speed	Comments
125–150 kHz (LF)	Basically unregulated	Â 10 cm	Low	Animal identification and factory data collection systems
13.56 MHz	ISM band, differing power levels and duty cycle	< 1 m	Low to moderate	Popular frequency for I.C. Cards (Smart Cards)
433 MHz	Non-specific Short Range Devices (SRD), Location Systems	1 – 100 m	Moderate	DoD Active
860–960 MHz	ISM band (Increasing use in other regions, differing power levels and duty cycle	1 – 2 m	Moderate to high	EAN.UCC GTAG, MH10.8.4 (RTI), AIAG B-11 (Tires), EPC (18000-6')
2450 MHz	ISM band, differing power levels and duty cycle	1 – 2 m	High	IEEE 802.11b, Bluetooth, CT, AIAG B-11

Fig. 1–8. RFID operating frequencies

The frequency range of 125–150 kHz is called *low frequency (LF)*; 13.56–433 MHz is *high frequency (HF)*; 300 MHz to 1 GHz is *ultra-high frequency (UHF)*; and 1 GHz and above is *microwave frequency.*

In Europe and certain parts of Asia, UHF RFID tags operate in the range of 865 MHz to 868 MHz. In the United States, the allowed range is 902 MHz to 928 MHz.

Radio frequency signals at the lower frequencies can travel through water and other liquids better than the radio frequency signals at higher frequencies. On the other hand, higher frequency RFID tags have a higher read range. Fig 1–9 shows the radio frequency properties of certain material types.

Material	LF	HF	ULF	Microwave
Clothing	RF-lucent	RF-lucent	RF-lucent	RF-lucent
Dry wood	RF-lucent	RF-lucent	RF-lucent	RF-absorbent
Water	RF-lucent	RF-lucent	RF-absorbent	RF-absorbent
Paper products	RF-lucent	RF-lucent	RF-lucent	RF-lucent
Motor oil	RF-lucent	RF-lucent	RF-lucent	RF-lucent
Metal	RF-lucent	RF-lucent	RF-opaque	RF-opaque
Wet wood	RF-lucent	RF-lucent	RF-absorbent	RF-absorbent
Plastic	RF-lucent	RF-lucent	RF-lucent	RF-lucent (certain types)

Fig. 1–9. Radio frequency properties of materials

The communication between an RFID reader and an RFID tag can be full-duplex or half-duplex. *Full-duplex* mode of communication implies that the reader and the tag can communicate simultaneously. In the case of the *half-duplex* mode of communication, the reader and the tag take turns communicating.

Read only or read/write tags

Read only tags have an identifier that is set during the time of manufacture and cannot be modified by the user at any time. Therefore, any other information that needs to be stored regarding the tagged item is stored in the database on the server.

A *writeable* tag is one where the tag can be written into by the user. This allows the user to write and rewrite data into the tag any time as needed. Some tags can also be *write-once*, which means the user can write only once after receiving it from the factory.

RFID tags vary significantly in their ability to store data. The amount of data that can be stored ranges from 1 bit to several kilobytes. As the storage capacity increases, so does the cost.

Kill or disable tags

Some tags have a provision where they can be permanently disabled if they receive a special code that instructs them to go into a "permanently disabled state." Once permanently disabled, they do not respond to read requests from RFID readers. This is often a safety measure to prevent unauthorized use or abuse of a tag. For example, if a consumer is buying an RFID-tagged item, for privacy reasons the consumer is probably not going to want the tag to remain active following the purchase of the item. Therefore, this feature can be used to disable the tag at the point where the item is purchased.

Security in tags

Some tags have the ability to communicate with readers using encrypted data. Encryption and authenticated data communication allow the tag data to be kept secure and accessible only to authorized users.

One of the most popular of such RFID tags is known as a *Digital Signature Transponder (DST)*. Manufactured by Texas Instruments, DSTs are deployed in several applications that are notable for wide-scale deployment and the high costs (financial and otherwise) of a large-scale security breach. These include:

- **Vehicle immobilizers.** More than 150 million vehicle immobilizer keys have been shipped with many newer automobiles, including 2005 model Fords, which use these Texas Instruments low-frequency RFID transponders.[2] This number includes systems with fixed-code transponders that provide no cryptographic security, as well as newer models equipped with DSTs. Immobilizers deter vehicle theft by interrogating an RFID transponder embedded in the ignition key as a condition of enabling the fuel-injection system of the vehicle. The devices have been credited with significant reductions in auto theft rates, as much as 90%.

- **Electronic payment.** DSTs are used in the ExxonMobil Speedpass system, with more than 7 million cryptographically-enabled keychain tags accepted at 10,000 locations worldwide.[3]

A DST consists of a small microchip and antenna coil encapsulated in a plastic or glass capsule. It is a passive device, which is to say that it does not contain an onboard source of power, but rather receives its power via electromagnetic inductance from the interrogation signal transmitted by the reading device. This design choice allows for a compact design and long transponder life.

A DST contains a secret, 40-bit cryptographic key that is field programmable using RF commands. In its interaction with a reader, a DST emits a factory-set (24-bit) identifier, and then authenticates itself by engaging in a challenge-response protocol. The reader initiates the protocol by transmitting a 40-bit challenge. The DST encrypts this challenge under its key, and truncating the resulting cipher-text, returns a 24-bit response. It is thus the secrecy of the key that ultimately protects the DST against cloning and simulation.

RFID Readers

RFID readers form the second layer of the RFID system stack shown in figure 1–10. A reader communicates with the tags to read them and write into them (if they are read/write tags) at the same frequency as that of the tags. The data read is passed on to the RFID middleware or directly to one or more applications (if RFID middleware is not used).

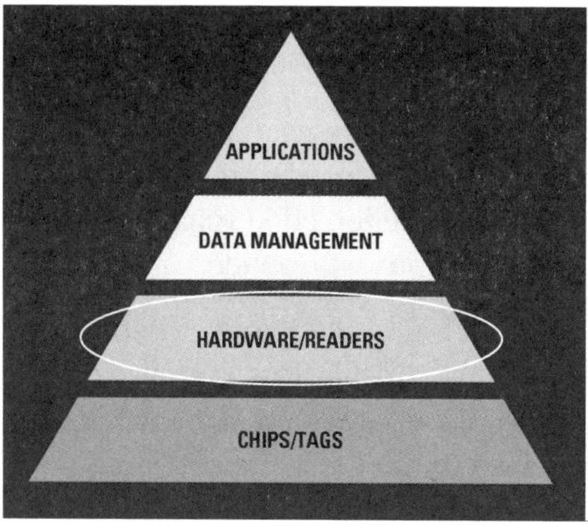

Fig. 1–10. RFID stack, focus on hardware/readers. *Source: Forrester Research*

As mentioned earlier, in the case of passive or semi-passive tags, the RFID reader transmits radio frequency signals to activate the tags that are within its range. Usually, a reader communicates only at a certain frequency with tags of the same frequency. However, readers are also available that can communicate with different types of tags at different frequencies. Such readers are referred to as *dual-frequency readers*.

Sometimes, readers have the capability of reading both RFID tags and bar code tags. This is particularly useful in the case of applications that use both bar codes and RFID-based identification. As mentioned earlier, bar codes are used for item-type identification whereas RFID allows the identification of unique items (i.e., the item type plus a serial number).

RFID readers can be *proprietary*, which means that they can read only the tags from the same manufacturer, or they can be *standards based*, which allows them to read the tags from different vendors. Obviously, the latter is more desirable.

RFID reader components

An RFID reader is comprised of the following hardware components:

- Antenna
- Microcontroller
- Network interface
- Power source

Antenna. The antenna of an RFID reader is the hardware that communicates with the antenna in a tag by transmitting and receiving radio frequency signals. Antennas can vary in size and range, depending on the type of reader and the read range that is desired (fig. 1–11).

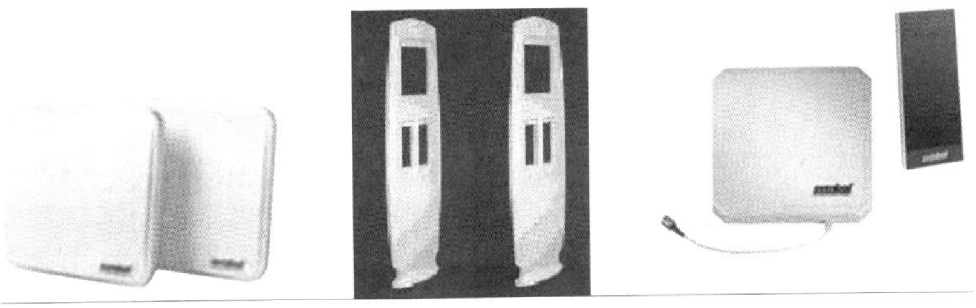

Fig. 1–11. Symbol Technologies antennas: (a) AN 200 general purpose antenna; (b) AN400 high performance area RFID antenna; and (c) AN480 single port RFID antenna

Some RFID readers have only one antenna that does the job of both transmitting and receiving radio frequency signals, while there are others that have separate transmit and receive antennas. There are also configurations in which an RFID reader has several antennas that are placed at different locations.

If an RFID reader has two antennas, one to transmit and one to receive, the positioning of the antennas becomes important (fig. 1–12). The antennas should be placed such that a moving tagged object comes within the range of the transmit antenna first and then the receive antenna. This allows the receive antenna to have more time to receive the data from the tag, thereby increasing the probability of a successful read.

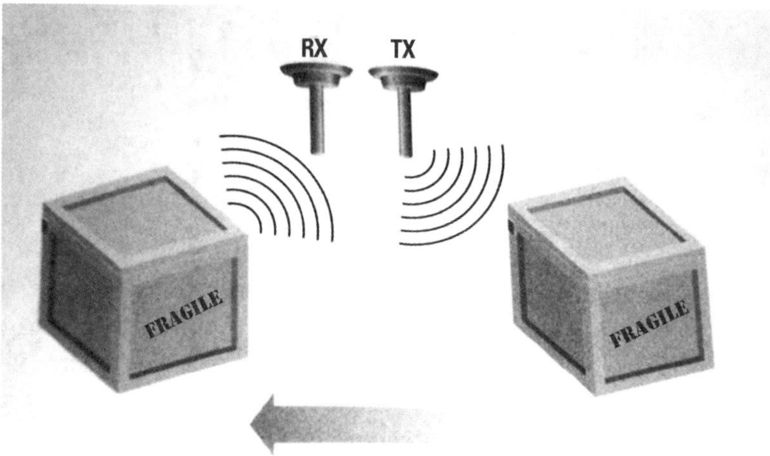

Fig. 1–12. Positioning of antennas. *Source: Figure 5.2: RFID Essentials, Bill Glover and Himanshu Bhatt, 2006*

In some cases a reader has multiple antennas, and each antenna can perform both transmit and receive functions, but they are placed at different locations. The antennas in this case may be considered to be either *logically one* or *logically separate*:

- **Logically one.** A configuration where all the antennas, though physically separate, are considered to be part of one logical antenna (fig. 1–13). When a tag is read by any of the physical antennas, it is considered to be part of the read by the same logical antenna and data is passed to the middleware layer accordingly. One example might be a warehouse door large enough to require an antenna on each side of the door (A-left and A-right) in order to ensure that the entire door is within the range of at least one antenna. A-left and A-right together comprise one logical antenna called *A*.

- **Logically separate.** A configuration where all the antennas that are connected to an RFID reader represent a different logical antenna. Consider a conveyor belt that has two antennas, A-1 and A-2, that are placed at locations L-1 and L-2, respectively, of the conveyor belt (fig. 1–14). A tag read by A-1 indicates that the tag is at location L-1, while a tag read by A-2 indicates that the tag is at location A-2.

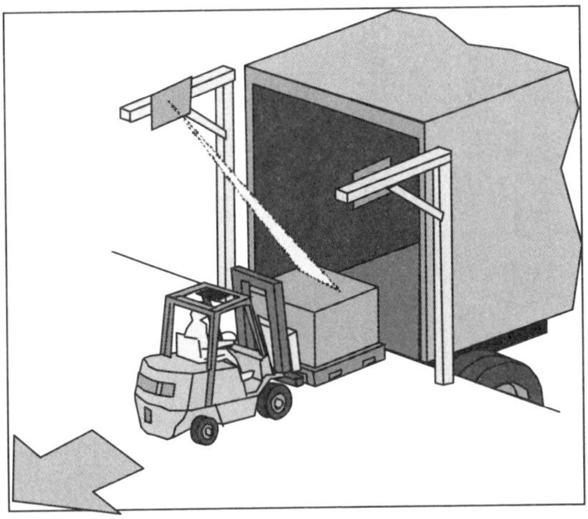

Fig. 1–13. Logically one antenna

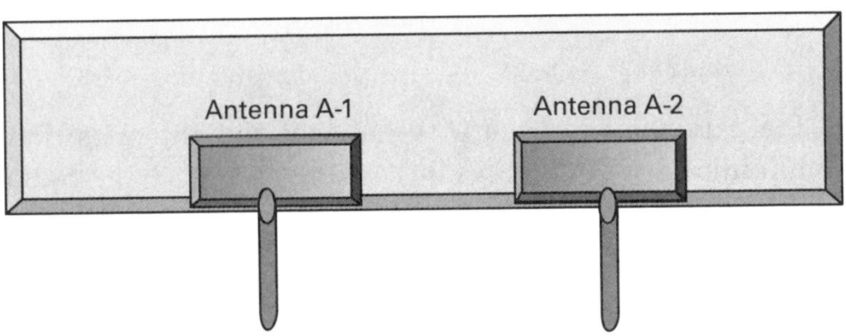

Fig. 1–14. Logically separate antennas for a conveyor belt

Microcontroller. The microcontroller is the processing unit of the RFID reader. The microcontroller interfaces with one or more antennas to communicate with the tags. The microcontroller also communicates with the networking hardware to forward the read RFID data to the RFID middleware or receiving applications that utilize the data for business processing (fig. 1–15). The microcontroller contains the processing software or the "brains" of the reader. Depending on the type of reader, the microcontroller may even have the capability of storing some of the data received from the readers.

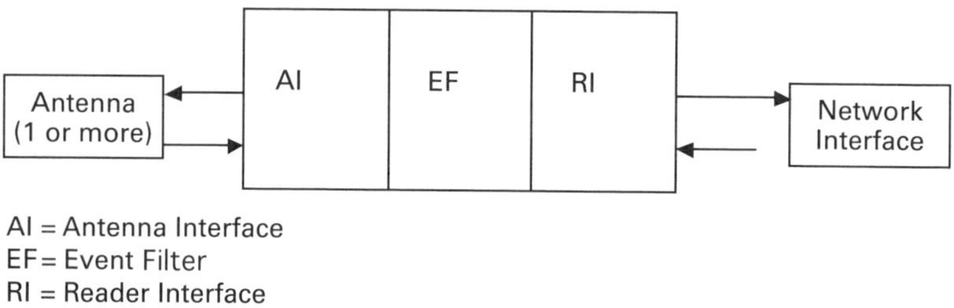

Fig. 1–15. Microcontroller software modules

The software in the microcontroller has the following subcomponents:

- Antenna Interface
- Event Filter
- Reader Interface

The Antenna Interface (AI) manages the antennas that communicate with the tags by sending and receiving radio frequency signals. It implements the necessary communication protocols to allow one or more antennas to communicate with the tags that are within range and to ensure that the responses from the tags are clearly picked up without collision. The protocol also allows the antennas to transmit data that needs to be written in a tag.

The AI is also responsible for configuring the antennas and monitoring their health to ensure that the antennas are functioning as per the requirements of the RFID system.

The data received by the AI from the antennas is passed to the event filter as raw data observations. An *event* is a change in the state of an observation (for example, a tag having an identifier of ID1 was not within the range of the reader previously but has just been observed to be within the range of the reader). The Event Filter layer identifies events from the raw data or observations and forwards the events to the reader API. This information thus can be made available to the RFID middleware or receiving applications (together referred to as *upper layer applications* or *ULA*).

The Reader Interface has the APIs that are used by the ULAs to communicate with the reader. The Reader Interface allows the ULAs to use this interface to configure the reader, monitor the health of the reader's components, register to receive alerts, and request event notifications.

Since RFID systems are designed to allow automated reading and writing of tags, it is essential to monitor a reader to ensure that it is functioning as per the requirements of the RFID system. Any failure or malfunction should be detected rapidly to prevent wrong conclusions from being made by the receiving applications that use the data for decision making. Active monitoring of a reader allows failure or malfunction to be detected rapidly. Also, if possible, the job of the malfunctioning reader can be handed over to another reader while the malfunctioning reader is being fixed.

Notifications involve the RFID reader sending information to a ULA when a certain event occurs. This is in response to a request that was made by the ULA.

An *alert* is a trigger generated by the reader to inform one or more ULAs about a certain condition in the reader or its environment to which the ULA needs to attend.

The implementation of the Reader Interface also includes the protocol implementation of the transport layer to allow communication over the network interface with the ULAs.

Network interface. The network interface allows an RFID reader to connect and communicate with the host computer or other data processing hardware, including a handheld device such as a PDA. Examples of such network interfaces are RS232 (serial port), RS422 (point to point, twisted pair), and RS485 (addressable, twisted pair). A reader can also come with a network interface controller (NIC) card to allow it to communicate with the host computer over the Ethernet. Recently, readers have also started supporting wireless technologies such as Bluetooth.

Power source. Readers also include a power source like a built-in battery or an external power source. In the case of handheld readers, if the reader is connected to a computer using a Universal Serial Bus (USB), the USB interface can provide it with 5 V DC power.

USB is a serial bus standard for interfacing devices that was developed to allow peripherals to be connected to a computer using a single standardized interface socket and to allow devices to be connected and disconnected without rebooting the computer.

Stationary or portable readers

RFID readers come in various shapes and sizes. The smallest ones can be embedded in PDAs (or mobile phones) and have an integrated antenna, microcontroller, and network interface.

Readers can be broadly classified as *stationary* (also called *portal*) or *portable*. In the case of stationary readers, the components of the reader, including any antennas, remain stationary while the tagged items move. The antennas are placed at fixed locations and configured to detect the tags within their range. Such readers can be deployed in warehouses or at docks, conveyor belts, and other locations. Tagged items entering or leaving a warehouse through a doorway can be detected by an arrangement of one or more antennas that are fixed and placed above the door or on either side of it (fig. 1–16).

Fig. 1–16. Symbol DC600 RFID slim portal system (includes Symbol XR400 reader, antenna, and cables)

Stationary readers can also be used in the case of smart shelves in a retail store or a storage facility. An array of antennas can be placed on the shelves in order to cover different parts of the shelves where tagged items are kept. The antennas have knowledge of all tagged items on the shelves, and this allows real-time inventory to be taken instantly at any time. Also, if a tagged item is added to a shelf or removed from it, that too can be automatically detected and the event can be passed along to the processing applications.

Portable (or handheld) readers are mobile readers that can be found as a separate handheld device similar to a handheld bar code reader or as a card that can be attached to a PDA or mobile phone. Some of these handheld readers not only look like bar code readers, they also include bar code reading functionality. Such readers are often rugged and are used in shop floors and warehouses (fig. 1–17).

Fig. 1–17. MC9090-G RFID mobile computer from Symbol Technologies

RFID readers can also be built into PDAs and mobile phones. There is currently an initiative called *Near Field Communication (NFC)* to have 13.56 MHz RFID reader functionality built into mobile phones and PDAs. NFC is discussed later in this chapter.

Other types of mobile readers are those that can be attached to forklifts and clamp trucks or mobile carts in warehouses and loading docks. These readers are designed for true mobility and therefore are cable free, have wireless capabilities, and include a battery and antenna (fig. 1–18).

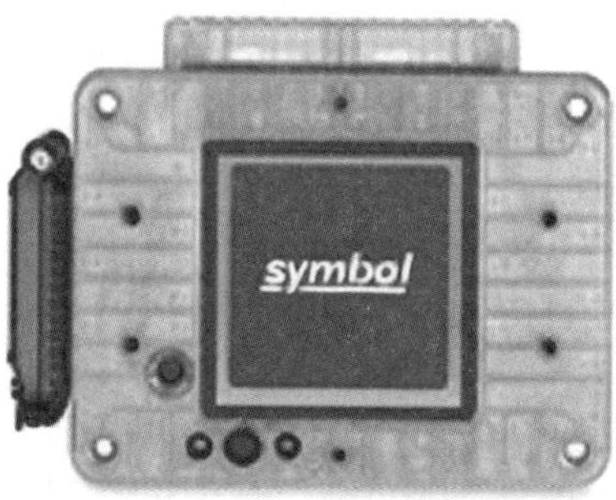

Fig. 1–18. RD5000 mobile RFID reader. *Source: Symbol Technologies*

The EPC reader protocol provides a uniform way for hosts to access and control the conforming readers produced by a variety of vendors. Different makes and models of readers will vary widely in the functionality they provide. Readers range from "dumb" readers, which do little more than report what tags are currently within the reader's RF field, to "smart" readers, which provide sophisticated filtering, smoothing, reporting, and other functionality. The reader protocol defines a particular collection of features that are commonly implemented and provides a standardized way of accessing and controlling those features when they are present. The EPC reader protocol does not require that all readers implement all such features, only that when such features are provided, they be made available to hosts in the manner specified herein.

Communication between Tags and RFID Readers

There are several protocols that facilitate communication between RFID readers and tags. The communication protocols can be broadly classified into two types: one where the reader initiates the communication first (*reader-first*) or the other where the tag initiates the communication first (*tag-first*). Passive or semi-passive tags and readers use reader-first, whereas in the case of active readers and tags, it can be reader-first or tag-first.

Tag-first works well where active reporting or alerting is needed. For example, an active tag might be equipped with a sensor that monitors the temperature of a machine in a plant. If the temperature exceeds a certain value, the tag can send out an alert signal.

In most cases, however, it is the reader that acts as the initiator. Slotted ALOHA, adaptive binary tree, and slotted terminal adaptive collection (STAC) are some of the most commonly used protocols where the reader initiates the communication.

When there are multiple tags within the range of a reader and they are communicating with the reader, their signals can interfere with each other. This interference phenomenon is called a *collision*. Protocols designed to prevent tags from interfering with each other while communicating with a reader are termed *anticollision protocols.*

The ALOHA protocol is one of the early networking protocols developed in the University of Hawaii in 1970. In the ALOHA protocol, as soon as a tag that is within a reader's range is energized by a reader, it sends its entire identifier (ID) and goes to sleep for a certain random duration (fig. 1–19). When it wakes up, it transmits its ID again. Even after receiving a tag's ID successfully, the reader does not send a response back. This protocol is simple but has a high probability of collision and therefore is not suitable if there are many tags in a reader's range.

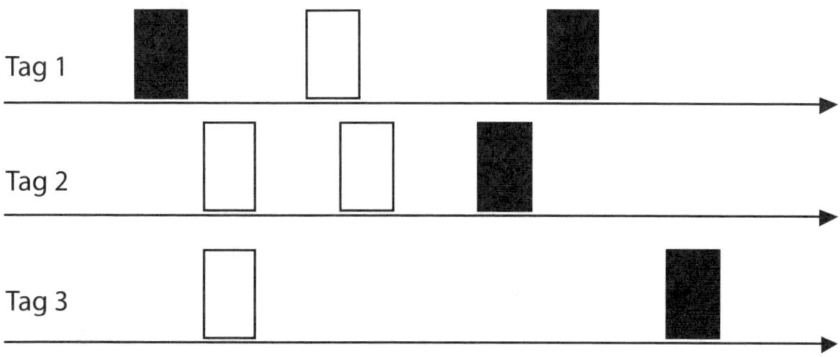

Fig. 1–19. ALOHA protocol. Black rectangle: successful transmission. White rectangle: collision

The slotted ALOHA protocol is a variation or refinement on the ALOHA protocol in which the RFID reader sends out a "wake-up" signal at the start of each cycle indicating the number of slots available for use by the tags within its range (fig. 1–20). A tag within the reader's range picks a slot by generating a random number and waits until it is time for it to transmit. It then transmits its ID and waits for an acknowledgement by the reader. If there is a collision because another tag also transmitted during the same slot, the tag does not receive an acknowledgement from the reader. In this protocol, packets either collide completely or do not collide at all. If there is a successful read by the reader, the tag enters a quiet stage and does not transmit any more until it is receives the next "wake-up" signal from the reader.

The slotted ALOHA protocol is designed to reduce the number of collisions compared to the ALOHA protocol, but it also has more protocol overheads. There are several variations of the slotted ALOHA protocol that have been designed to further reduce the probability of a collision.

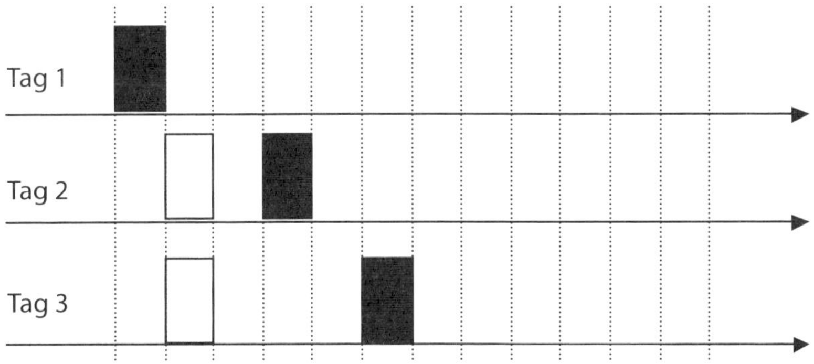

Fig. 1–20. Slotted ALOHA protocol

RFID Printers

An RFID printer is a device that can encode an RFID tag and also print on the label packaging of the tag (fig. 1–21). An RFID printer has the following components:

- Reader
- Verifier
- Printer
- Applicator

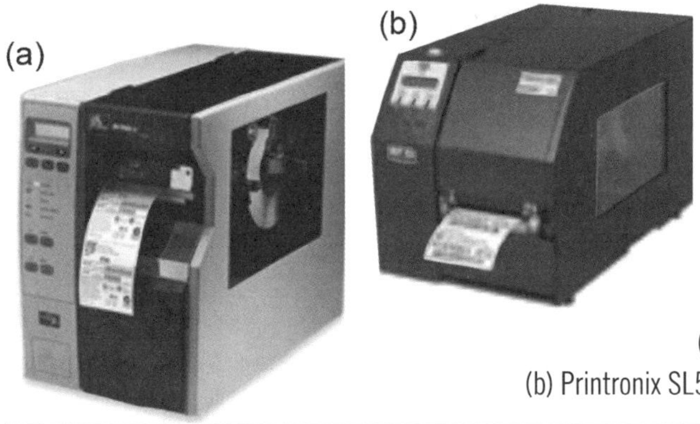

Fig. 1–21.
RFID printers
(a) Zebra R110Xi
(b) Printronix SL5000r MP2 RFID

The reader component has the same capability as that of a typical RFID reader. It encodes and writes into an RFID tag as an RFID reader, while the printer component prints text or a bar code, or both, on the paper label that encapsulates the tag. The printed text includes the tag identifier and other details on the product. The verifier is responsible for verifying whether the tag is functioning properly (i.e., the data written could be read correctly by the reader). Verification is usually done by the same reader. Sometimes, though, external verification by a separate RFID reader may be required after the applicator has applied the tag on an item.

The printer component has capabilities similar to those of laser, inkjet, or thermal printers.

RFID Middleware

The RFID middleware is software that resides between the RFID readers and the receiving applications. It receives RFID data (or observations) from one or more RFID readers and converts them into events that are meaningful to the receiving applications. It then forwards the events to the receiving applications, as requested.

Figure 1–22 shows the RFID middleware receiving RFID observations from readers R1, R2, R3, and R4. On the other side, it communicates with four receiving applications: RA1, RA2, RA3, and RA4. The middleware does the job of brokering data between the readers and the receiving applications.

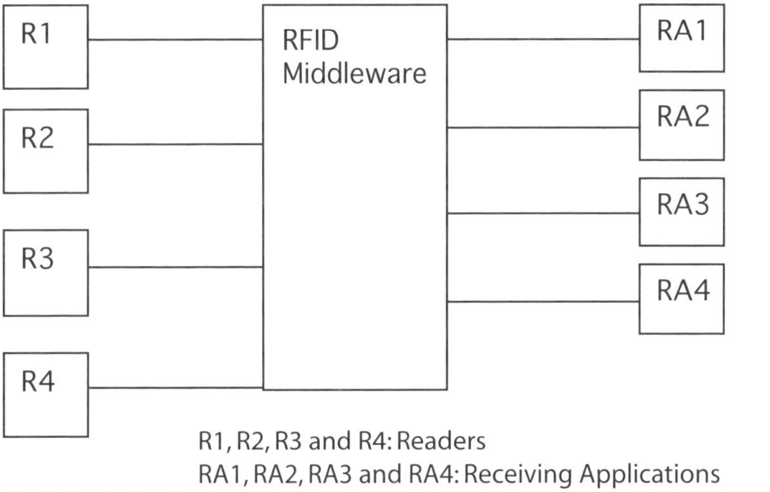

Fig. 1–22. RFID middleware with readers and receiving applications

Figure 1–23 shows the logical components of the RFID middleware software. The reader adapter is responsible for communicating with RFID readers. Since the readers talking to the middleware can be from multiple vendors and have different interfaces, the reader adapter abstracts this functionality from the higher level applications to expose a single interface. This eliminates the need for the receiving applications to deal with the reader interfaces of different RFID readers.

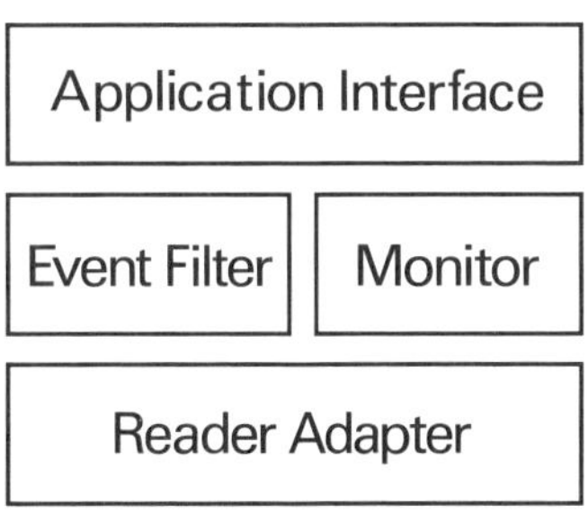

Fig. 1–23. RFID middleware components

The event filter receives the raw observations from the readers and converts them into events that are meaningful to the receiving applications. It provides application level context to the RFID data. Additionally, by converting the observations into events, the event filter also significantly reduces the amount of data that a receiving application needs to process.

The *monitor module* is responsible for the monitoring and management of the readers that are connected to the RFID middleware. Since RFID systems provide a higher level of automation compared to other technologies such as bar code, automatic detection of any failure in RFID readers is essential. Failure or malfunction in readers results in notification being sent out to appropriate personnel. Also, it is desirable that the work of the failed reader be assigned to one or more readers during the interim period until the malfunctioning reader is fixed or replaced and is fully operational again.

The application interface is the interface for the receiving applications to communicate with the RFID middleware. The communication with the receiving applications can occur in *synchronous mode* or *asynchronous mode* (figs. 1–24 and 1–25). In synchronous communication a request is made by a receiving application to the middleware, which results in the middleware sending a response back to the receiving application synchronously. The request-response mode of communication results in higher bandwidth usage compared to asynchronous communication.

In the case of asynchronous communication, the receiving applications subscribe to one or more events with the RFID middleware. When an event occurs, it is published by the RFID middleware such that all of the receiving applications that had subscribed to the event receive a copy of it. This publish-subscribe communication model is an asynchronous messaging paradigm where the sender of an event is not aware of the subscribers to the event. Subscribers express interest in one or more events or classes of events and receive only those events in which they had expressed an interest. This mode of communication is more efficient than that of request-response, since the receiving application does not have to keep polling the RFID middleware for events.

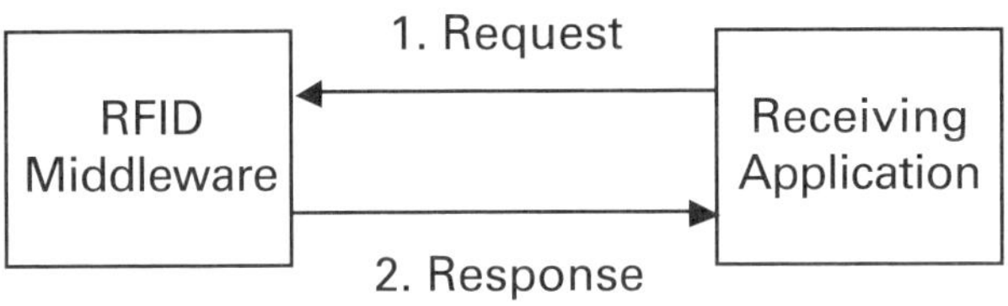

Fig. 1–24. Synchronous communication

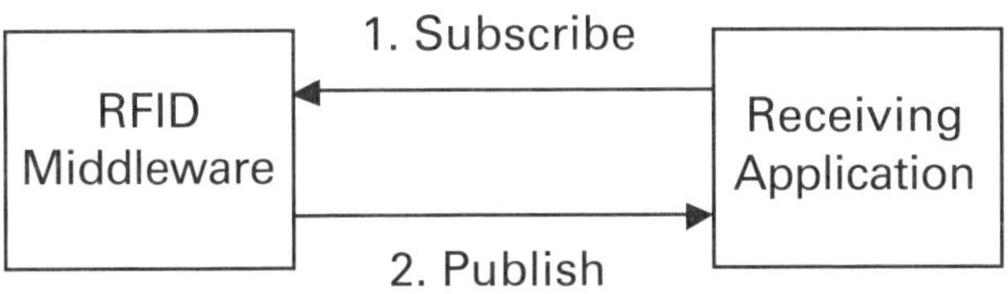

Fig. 1–25. Asynchronous communication

Thus, the RFID middleware plays an essential role in facilitating the data exchange between the RFID readers and the receiving applications by making the data more relevant to the business applications.

Applications

The applications layer comprises one or more receiving applications that utilize the RFID events generated by the RFID system to provide visibility into various business processes. The RFID readers in the RFID system convert the raw data observations into RFID events, which are then passed on to the middleware layer. The middleware receives RFID events from various readers and further processes the data to generate RFID events that are meaningful to the receiving applications. The raw observations are not directly meaningful to a receiving application.

A receiving application processes the RFID event data along with the business context data to generate reports that provide visibility into business processes. Business context data is comprised of data specific to the nature of the application that provides the meaning or context to the RFID events. For example, in a warehouse application, business context data can be comprised of the list of products and detailed information on each product. For a receiving application that is used to manage the fixed assets (such as a valve or switch) in a water utility plant, business context data can be comprised of the maintenance and repair history of assets.

An RFID event has the following data elements:

- **Identifier.** The identifier of the RFID tag that resulted in the event.
- **Time stamp.** The time when the event was generated. If the time stamp is inserted by the RFID reader, the clocks of all the readers must be synchronized. Consider an example in which a tag was read by reader A at time Ta, and then by reader B at time Tb. If the clocks of readers A and B are not synchronized, then even if Ta < Tb, it is not necessarily correct to conclude that the tag was read by reader A first and then by reader B.
- **Location.** The location where the RFID tag was read.

In addition, an event may include other data as well. For example, a tag placed inside a pump station may also record the temperature using a sensor. The event in that case may include the temperature reading also, as an additional data element. If a sensor is integrated with a tag, it is possible to measure temperature, current, pressure, and other such attributes that can then be passed along as data elements in an event.

Figure 1–26 outlines the components that a typical receiving application can be expected to have.

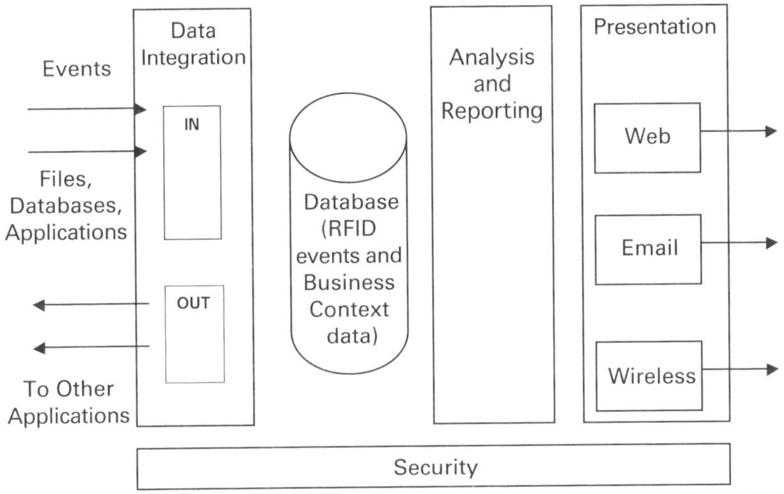

Fig. 1–26. Receiving application components

Data integration

The data integration module enables data to be input into the receiving application, as well as to be channeled out to other applications. In figure 1–26, *IN* represents the module that is responsible for inputting data into the database, while *OUT* represents the module that is responsible for channeling data out to other applications.

The inputs to IN can be RFID events from the RFID middleware, as well as data from files, other databases, or other applications. Files can have data in structured or unstructured form. Examples of structured data files are CSV (comma separated values) files and XML files.

Unstructured data is defined loosely as any data that does not fit neatly into rows and columns within a database or spreadsheet. This includes most of the files that one encounters daily, including e-mail, Word documents, PDF documents, and Web pages.

The input data received by the IN module is stored in the database.

The OUT module is responsible for channeling data out of the receiving application to other applications. It does this by generating the output data as CSV, XML, or in another format that can be understood by the other application. The output data is handed over to the other application using TCP/IP, HTTP, Web services, or some other communication protocol.

Database

The events and business context data are stored in the database. A relational database such as Oracle is often used. One will need to calculate the database size depending on the amount of RFID data that is generated by the system and the length of time the data needs to be kept in the database. This is an important consideration since the volume of RFID data generated can be overwhelming.

Another important consideration is to determine the archival strategy. Archiving is the process of moving data into a secondary storage that can be accessed readily whenever required. The RFID event data should be archived after a certain time.

Analysis and reporting

This module analyzes the event data and business context data as per the business rules specified to generate reports. The business rules are instructions specified by a business user on how the event data and its associated business context data need to be processed together to generate the reports.

Typically in an organization, different types of reports need to be generated so that managers, depending on their responsibility, can get access to the right information. For example, a senior executive who has the responsibility of the entire department would require a higher level view indicating how the overall process is functioning. A supervisor might need an update every hour with details on any issues or problems that may occur.

A good example is the frequent occurrence of misshipments for the fast-moving consumer goods industry. Some of the misshipments are valued in millions of dollars, and returning them to the third-party logistics provider in all cases is not the most prudent action. In such cases the RFID infrastructure can provide needed exception alerts to the senior executives, accept their input via a PDA, and close the loop with dock workers in real time.

It is important to have a system where the business rules can be easily added or modified by the business users, instead of requiring the IT department to write code every time a new requirement comes up. This provides flexibility to the business by enabling the business users to obtain the right visibility into a process based on the current circumstances or situation.

Presentation

The presentation module allows information to be presented or displayed to business users via Web-based tools, e-mail, or wireless mobile devices. Information that is made available using the presentation module includes the following:

- **Reports.**
- **Notifications.** Notifications inform the business user of an event.
- **Alerts.** Alerts are special notifications that are generated to inform the business user of an event as soon as it occurs. An alert can often involve triggering an alarm, sending out a page, or even giving a visual notification to an operator.

It is important to provide a user interface that is graphical and interactive to allow the user to make use of the information rapidly, without having to be "technologically oriented."

Security

The security module is responsible for ensuring that only authorized users are able to use the application and access the data. Authentication, access control, and encryption technologies required for an RFID system are detailed below.

The receiving applications may require other software such as Web application servers, relational databases, a wireless gateway, system management software, and even advanced reporting software. A variety of software from third-party vendors is available in the market today. An example is provided in appendix A.

Security and Privacy

With any technology, security and privacy are important considerations. Only those with the appropriate permissions should be allowed access to the RFID system. Intruders should not able to break into the system to read the data or, even worse, modify it. In an RFID system data is present in the tags, the readers, and the applications in the host computer. Data is also vulnerable when it is transmitted from the tags to the readers or exchanged between the readers and the applications wirelessly or in a wired network.

At every stage, it is essential that the data is not compromised to unauthorized personnel since the consequences can often be severe. The type of security required depends on the nature of the application and the type of data that the application deals with. For example, the security requirements of an RFID system that tracks items in a warehouse will be different from one that involves consumers using RFID-enabled access cards to enter a residential or office complex.

Any time one considers a system that is used by consumers, any compromise in data can put them at risk. Such risks involve knowing what the consumer buys, the consumer's health, or even identity theft. In extreme situations it can result in a threat to the consumer's life.

For example, consider an RFID system that is used to locate, maintain, and monitor fixed assets in an airport or a power plant. RFID-based smart labels can be placed at different locations, either on or in the vicinity of an asset so they can be used to assist an operator during maintenance operations or an emergency. If an intruder such as a terrorist can read these labels, it can be a serious threat to the national security of the country.

As discussed earlier, an RFID reader can read an RFID tag without requiring a direct line of sight. This is more advantageous than a bar code since the level of automation can be much higher. On the other hand, compromise in security as a result of an illegitimate read or write of a tag performed by an intruder is much harder to detect. Therefore, as with every new technology, the benefits and risks of the system need to be carefully evaluated before undertaking a project.

The security threat posed by a read-only tag is no different than that of a bar code or a tag with a printed serial number written on it.

Privacy issues

Privacy is the ability of an individual or group of people to prevent information about themselves to be known to people other than those to whom they are agreeable to provide the information. This allows the individual or the group to hold on to their personal information without any imposition by others.

Whenever a technology like RFID is used in an application that involves consumers, there is often the possibility of it encroaching on their privacy. For example, if RFID technology is used to tag items that are sold in a retail store, when a consumer makes a purchase, the association of the item with the consumer can result in privacy issues. If the tag attached to the item is not disabled, it accompanies the consumer. Subsequently, if the tag is read by a reader, it can be used to derive information on the consumer's purchase habits, financial capabilities, or health. The consumer may be buying certain products or drugs to address a certain health condition and would very much like for it to be private.

If thieves are able to get such information, it can help them determine which houses to target and when.

Another example could be considered. If a mass transit system uses near field communication (NFC) or other RFID technology for ticketing, it can allow passengers to use RFID-based smart cards or their mobile phone as tickets. While this is very convenient for passengers, it also allows the transit system management to know when and where a certain passenger boards the train. This information in the hands of the wrong people can be a serious security threat to that individual.

It is true that privacy becomes a serious consideration when using RFID-based tags, but in many ways such privacy issues already exist today even without RFID. With RFID, since a line of sight is not needed for a read or write, these operations may be performed more easily without the knowledge of the consumer. However, if the read/write range is small (as in the case of 13.56 MHz tags), this becomes less of a threat.

One way to prevent this threat is to "kill" a tag (see below) or disable it permanently once its intended use is over.

Security considerations of an RFID system

An RFID system typically includes tags, readers, PDA or other mobile devices, computers, and networking hardware. The data exchange between these components can occur wirelessly or over a wired network. Wireless data transfer technologies possible include wireless LAN, wireless WAN, Bluetooth, and others. The data transfer between a tag and a reader is also done wirelessly.

The following section looks into some of the security considerations of the RFID system in more detail.

Physical security. Any of the hardware elements that are a part of the RFID system should be prevented from theft. Whether the system has to do with mobile asset tracking or applying the RFID tags on fixed assets as smart labels, it is crucial to prevent the tags from being stolen or damaged.

The same applies for the other hardware components of the RFID system—RFID readers, PDAs and other mobile devices, or computers. All hardware should be physically secured against loss, theft, or damage. For example, computers need to be in a data center where access is restricted to authorized staff who would need an electronic access card to enter the facility. Also, all valuable assets in the facility can be tagged using RFID tags so if any such asset is removed from the facility, it is always a known event. Unauthorized access or an attempt to steal an asset results in an alert to security personnel.

In addition to the assets that are a part of the RFID system, other assets that indirectly ensure smooth operations of the RFID system also need to be secured and protected. Examples include power supply and air conditioning systems in a data center.

Malicious attacks and threats. RFID systems are also susceptible to attacks such as denial of service, spoofing, and eavesdropping.

A *denial-of-service attack* is characterized by an explicit attempt by attackers to prevent legitimate users of the RFID system from using it. Such attacks include attempts to flood a part of the network, disrupt the connections between two or more components of the system, or prevent a particular individual or application from using the system properly. Denial of service may also involve jamming the radio frequency signals between tags and readers.

A *spoofing attack* is a situation in which an application, tag, or reader successfully masquerades as another by falsifying data thereby fooling the system to believe something that is not true. For example, by spoofing valid tags an attacker could fool the RFID system into believing that a product is still on a warehouse shelf when it is not. Alternately, someone could modify the data in a writeable RFID tag that would make the RFID system result in incorrect conclusions.

Eavesdropping is the act of listening in on or wiretapping data when it is exchanged between an RFID tag and an RFID reader, an RFID reader and the RFID middleware, or other components of the RFID system.

User authentication and encryption are techniques that are used to prevent some of these attacks.

Authentication. *Authentication* is technology that allows a user to access the RFID system only after the user's identity is verified as someone with the appropriate permissions. This is done by requiring the user to sign in with a valid username and password. The password is the secret code that should be known only to the user having the particular username. It is important to use a password that cannot be guessed easily by an unauthorized person who has somehow obtained the username of the user. A random sequence of letters and numbers of at least eight characters is recommended to prevent it from being easily deciphered by others.

Encryption. *Encryption* is the process of transforming data to make it unreadable to anyone except those who have the *key*, a special code. Encryption has long been used by the military and governments for secret communication.

Encryption can protect the confidentially of the data, but other techniques are still needed to verify its integrity and authenticity. Several standards for encryption such as DES and public key encryption are now widely used. The *Data Encryption Standard (DES)* uses 64-bit or 128-bit keys. *Public key encryption* is a cryptographic technique that involves two keys—a public key and a private key. The public key is distributed to others while the private key is kept secret. A message encrypted with the public key can be decrypted only with the corresponding private key. Even though the keys are mathematically related, the private key cannot be derived using the public key.

If the data stored in an RFID tag is not encrypted, it can be read by any RFID reader that operates at the same frequency. If it is a writeable tag, the data in the tag can be maliciously modified. Similarly, when data is stored in the RFID system, either in a file or database, it should be encrypted to prevent an unauthorized read. Encryption is also needed when data is transferred over a wired or wireless network.

Firewall. A *firewall* is a hardware device or software that can be used to permit or deny data connections into a private network as per the configured security policy of the organization (fig. 1–27). Basically, it controls traffic between networks with different zones of trust, such as between the Internet (no trust) and a private network of an organization (meant to have a high level of trust). It restricts access to the private network to only a certain set of functionality.

All computers that form a part of the RFID system to store and process data should be protected with a firewall.

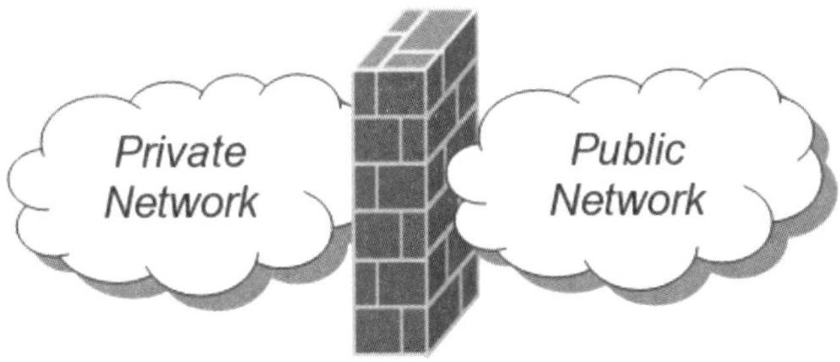

Fig. 1–27. Firewall

Disabling a tag. As mentioned earlier, it is often desirable to disable a tag after its intended use is over. A consumer may not want the RFID tag on an item to be readable after he or she has bought the item. An RFID tag on a medicine bottle needs to be disabled permanently to prevent counterfeiters from reusing the bottle. The simplest way to disable the tag is to break it. Another way is to electronically disable it by using a *kill* command, a special command that when received by the tag with a certain password, results in the tag disabling itself permanently.

Manageability and System Availability

An RFID system as described above is comprised of RFID tags that are attached to objects, one or more fixed RFID readers placed at different locations (or portable RFID readers) that are used to read (and sometimes write) the RFID tags, and one or more host computers. The host computer is comprised of one or more servers, as well as other computers that may be placed at intermediate locations to receive data from the readers. The host computer runs the RFID middleware and receiving applications to utilize the RFID events for decision making.

In figure 1–28, reader A is placed in location A, reader B in location B, and readers C and C´ in location C. The output observations from these readers are channeled to the RFID middleware, which does event filtering to convert the observations into events and then forwards the data to the receiving applications.

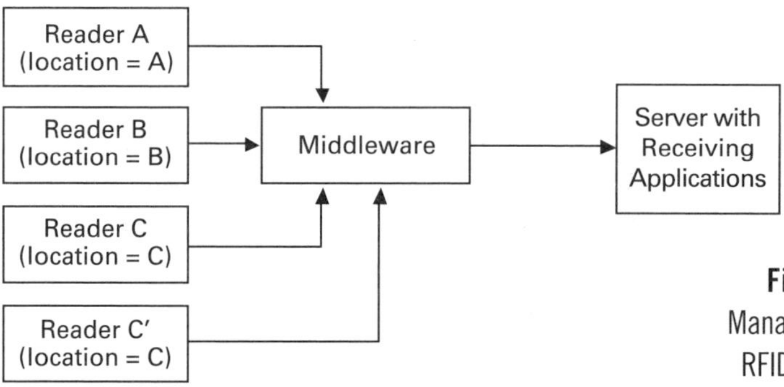

Fig. 1–28. Managing the RFID system

Since RFID systems are designed for automated reading, a higher level of automation is necessary in managing the hardware and software components. Automated system management allows failure to be detected rapidly and, if possible until the failed resource is replaced, to distribute the tasks of the failed component among other resources of the same type to prevent or reduce downtime.

Some of the capabilities of a system that provides automated system management are as follows:

- **Automated detection and configuration.** When a new hardware device or software is added to the RFID system, it should be able to automatically detect the hardware and software resources that it needs, such as RFID readers or RFID middleware. Then it should be able to obtain its configuration information from the management software and accordingly configure itself. The management software for an RFID reader could be the RFID middleware.
- **Remote management.** The hardware and software resources of the RFID system should be manageable by "remote management" software that communicates with them to periodically monitor their health. If the remote management software detects a problem or failure with a resource, it takes remedial actions such as the following:
    - *Notifying operators by e-mail, fax, SMS, or page informing them of the problem.* Depending on the location of the fault, notifications are usually sent to the appropriate operators in order that the problem or failure can be attended to rapidly. In addition, a log is also made at the central station or location.
    - *Determine if the functionality or tasks of the failed resource can be performed by another resource during the interim.* For example,

in figure 1–28, if reader C fails, it may be possible for reader C' to take over the job of reader C until it is repaired or replaced. Such actions enable the system to operate without a downtime.

- **Software or firmware upgrade.** It is important that the software and firmware used by the resources be easily upgraded. Automated software upgrades allow the entire system to be kept updated with the latest compatible versions of software, without having to go through manual configuration. The resources should be able to detect the presence of newer versions of software and upgrade by downloading and installing automatically.

Simple network management protocol (SNMP) is one of the popular protocols in use today for managing devices in a network. It allows an operator to query the attributes of a resource using "Get" calls and assign values to the attributes using "Set" calls. When a resource wants to send out a notification to the management software, it sends out a "Trap."

NFC Technology

Near field communication (NFC) is a short-range wireless technology mainly aimed at usage in mobile phones and other mobile devices. The technology is one of the newest wireless technologies that evolved from existing contactless identification and interconnection technologies.

NFC operates at 13.56 MHz radio frequency by magnetic field induction. Using NFC, it is possible to transfer data at rates of up to 424 kbits/sec between two NFC-enabled devices. NFC is both a read and a write technology. Communication between two NFC-compatible devices occurs when they are brought within 0 to 4 centimeters of one another. Given this short transmission range, NFC-enabled communications are inherently secure.

The technology used is passive RFID at 13.56 MHz. The initiator device generates the radio frequency field that actives the target device, which responds with its identifier and other data. The target device draws its operating power from the radio frequency signal. NFC can be used to configure and initiate other wireless network connections such as Bluetooth or Wi-Fi. The underlying layers of NFC technology follow universally implemented ISO, ECMA, and ETSI standards.

Some of the key applications of NFC are as follows:

- Travel—electronic ticketing and travel cards
- Provide location-based information
- Asset tagging and identification
- Smart wallet and mobile commerce
- Electronic keys
- Simplify wireless communication and connectivity

Location-based applications can be used by utilities, the oil and gas industry, power plants, and other large facilities to locate, maintain, and monitor their fixed assets such as valves, switches, and underground pipes.

Figure 1–29 shows applications of NFC technology. NFC technology is expected to enable the adoption of other wireless technologies such as Bluetooth, Wi-Fi, and ZigBee among nontechnical users. For additional information on NFC, one can go to the Web site at www.nfc-forum.org.

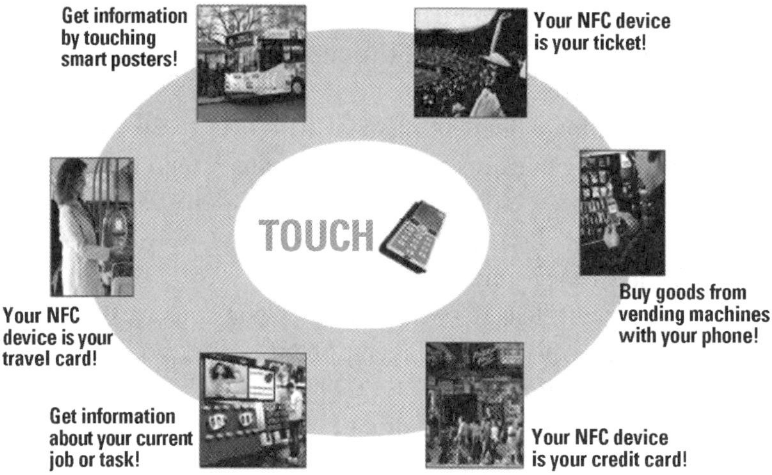

Fig. 1–29. Applications of NFC. The possibilities for using Near Field Communication are nearly limitless. The potent attraction of touchless transactions will help weave NFC technology into the fabric of our daily lives. *Source: NFC Forum's Marketing White Paper*

References

1. ANSI/IEEE. 2000. Recommended practice for architectural description of software-intensive systems. Discussion and FAQs accessible at http://www.iso-architecture.org/ieee-1471/ieee-1471-faq.html.

2. Texas Instruments. 1999. Automotive immobilizer anti-theft systems experience rapid growth in 1999. Press release. (June 1). http://www.ti.com/tiris/docs/news/news releases/90s/rel06-01-99.shtml.

3. ExxonMobil. 2004. Speedpass press kit fact sheet. (June). http://www.exxonmobil.com/corporate/files/corporate/speedpass_fact_sheet.pdf.

4. Glover, Bill, and Himanshu Bhatt. 2006. RFID essentials. O'Reilly. (January).

5. NIST Special Publication 800-98. 2007. *Guidelines for Securing Radio Frequency Identification (RFID)* Systems. (April).

2 WIRELESS NETWORKING TECHNOLOGIES

Introduction

In this chapter, various wireless networking technologies that are of relevance in deploying RFID-based applications are examined. These networking technologies have evolved over the past few years and have become popular in networking devices and computers in a variety of areas such as personal networks, local area networks, and wide area networks. Although the most popular application of wireless communications today is mobile telephony, this discussion will be restricted to wireless technologies that are used for data networking.

Categorization of wireless networks

There are four categories of wireless networks based on the coverage area over which these networks are expected to operate. The expected coverage area or the maximum distance between communicating wireless networking devices determines many important design parameters for a wireless technology. Design parameters include transmission frequency, transmission power, modulation scheme, collision avoidance, and antenna shape and size. The four main categories of wireless networks are the following:

- Wireless Personal Area Network (WPAN)
- Wireless Local Area Network (WLAN)

- Wireless Metropolitan Area Network (WMAN)
- Wireless Wide Area Network (WWAN)

Wireless personal area network (WPAN). A WPAN is expected to operate in the personal space around a human being. These networks are typically operated over short distances (e.g., less than 10 m) and can be used to interconnect personal devices such as personal computers (PCs) and mobile phones. Some applications of WPAN include elimination of wires in connecting peripherals such as headphones to PCs and mobile phones, and wireless synchronization of personal devices such as a personal digital assistant (PDA) and a PC.

Wireless local area network (WLAN). WLANs operate over areas ranging from a few meters to a few hundred meters. WLANs were designed primarily to connect mobile/portable computers such as laptops to local area networks (LANs) within enterprises. Today however, WLANs are used for a variety of applications, the most popular of which is Internet access for travelers at the airports, coffee shops, train stations, and even in some city centers. Outdoor deployments of WLAN mesh networks can also provide citywide coverage for emergency services such as police and fire departments.

Wireless metropolitan area network (WMAN). WMANs typically operate over distances from a few hundred meters to a few kilometers and are thus ideal for blanketing a city or a metropolitan area with wireless networking coverage using significantly fewer base stations/access points than is possible using WLAN. Typical applications of WMAN include broadband Internet access for residences and businesses (wireless replacement of DSL and cable) as well as point-to-point communication between buildings in a city or between communication towers. Recently, WMANs have also been designed for high-speed networking for mobile data terminals at pedestrian and even vehicle speeds.

Wireless wide area network (WWAN). WWANs have evolved from traditional mobile telephone networks to also provide data networking to mobile devices such as smart phones, PDAs, and even laptop computers. WWANs also typically support high-speed mobility over a wide geographic area such as a state or even a country using *cells*, or areas covered by a base station. When a mobile terminal moves from the area covered by one base station to another base station, a *handoff* occurs between the two base stations as the mobile terminal changes its point of wireless access from one base station to the other. WWANs use the same infrastructure as mobile telephone networks and are usually limited in their data speeds compared to WMANs. This data speed limitation occurs because WWANs were designed primarily for voice, whereas WMANs were designed primarily for high-speed data networking applications.

Evolution of wireless networks

One of the earliest nonmilitary instances of wireless communications for data networking was the experiment to link University of Hawaii campuses from different Hawaiian Islands in 1971.[1] This experiment resulted in many of the protocols such as ALOHA that formed the basis for today's advanced communication technologies including Ethernet and wireless LANs.[2] Later, in 1990, NCR (later acquired by AT&T) came up with the first commercial implementation of wireless LAN called WaveLAN. This equipment operated in the 900 MHz industrial, scientific, and medical (ISM) band, which was earlier freed up by the Federal Communications Commission (FCC) in the United States for unlicensed use by wireless applications. Maximum data rate possible with WaveLAN was 2 megabits per second (Mbps). Also in 1990, the Institute of Electrical and Electronics Engineers (IEEE) formed an 802.11 Working Group to standardize wireless LANs. The first products based on the early 802.11 standard were released by AT&T under the name of IEEE WaveLAN in 1997. In 1999, a faster variant known as 802.11b was standardized to operate at 11 Mbps in the 2.4 gigahertz (GHz) ISM band. Other variants known as 802.11a and 802.11g were standardized in 1999 and 2003, respectively, to increase the speeds to 54 Mbps.

In parallel with WLAN standardization efforts of IEEE, mobile cellular telephony was also getting established in developed countries. One of the earliest implementations of cellular data networking was known as *cellular digital packet data (CDPD).* CDPD operated over traditional advanced mobile phone system (AMPS) cellular networks in the 800 MHz and 900 MHz licensed frequency bands at speeds of up to 19.2 kbps and required separate modems. With the emergence of second generation (2G) digital cellular networks such as time division multiple access (TDMA), global system for mobile communications (GSM), and code division multiple access (CDMA) in late 1990s, data access at rates of 9.6–14.4 kbps and higher was part of the mobile telephony standards. Third generation (3G) cellular networks are already being deployed in several developed countries and promise data rates of 1 Mbps or more.[3]

WPANs came into existence largely with the need to replace wires in short-distance networking applications. These applications include connecting peripherals such as headphones, keyboard, and mouse to desktop and laptop computers, as well as interconnection of cell phones with personal computers. In 1998, the Bluetooth Special Interest Group (SIG) was formed to standardize short-range wireless technology, and the first Bluetooth products were available in 2000.[4] Also, IEEE established the 802.15 Working Group in 1999 to standardize wireless personal area networks (WPANs). More recently, ultra-wideband (UWB) technology has emerged for WPAN promising high data rates in excess of 100 Mbps at short distances. In addition, ZigBee is another WPAN technology popularly used for low-power and low-data-rate sensor network applications.[5]

WMANs were developed around 1999 to 2000 to provide fixed broadband wireless access (BWA) to residential and business customers using microwave frequencies. The BWA market still exists in developing countries where wired networking infrastructure is not widely deployed. In more developed countries however, WMAN deployments are now targeting mobile broadband data services with Worldwide Interoperability for Microwave Access (WiMAX) as the dominant technology.[6]

Each of the four network categories will now be examined in detail, including basic architecture, components, and typical applications.

Wireless Personal Area Networks (WPANs)

WPANs were developed primarily as a technology for cable replacement in connecting personal devices to other devices and peripherals. Initial applications included cordless headsets and earbuds for hands-free operation of mobile phones, cordless keyboards, and mice for PCs, in addition to wireless synchronization between two mobile phones or a mobile phone and a PC. For use in small consumer electronics devices, many of which are battery operated, a WPAN component needs to be small in size, low power, and low cost. Other requirements include sufficient data rates for intended applications and robustness against interference from other nearby WPANs as well as other wireless technologies. WPANs need to be largely self-configurable so that WPAN devices can communicate with other devices in their vicinity, forming ad hoc networks without requiring much user intervention. Because of all these properties, WPAN technologies have also found applications in sensor networks.

Architecture and components

WPANs are organized as *piconets*, each one being an ad hoc collection of devices. Each piconet is typically controlled by a *master* or a WPAN *coordinator*. Multiple piconets can coexist in a single physical space. Also, a single device can be part of multiple piconets. Figure 2–1 shows typical usage scenarios for WPANs, while figure 2–2 shows some examples of piconets.

Examples of piconet topologies include a *star*, a *mesh*, and a *cluster tree*.[7] Full function devices can forward packets from other nodes, whereas reduced function devices can only send or receive packets for themselves.

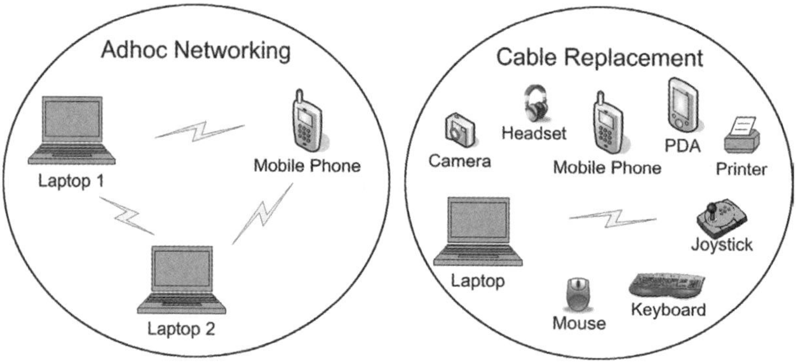

Fig. 2–1. Typical usage scenarios of wireless personal area networks

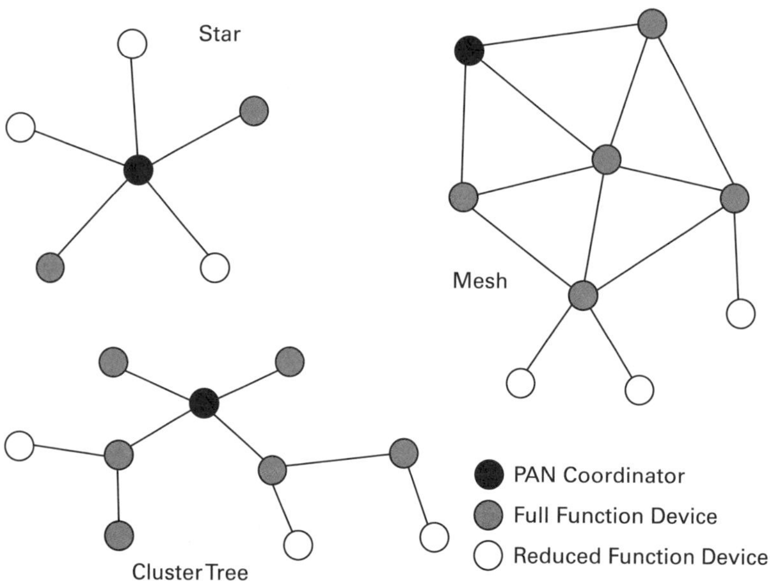

Fig. 2–2. Examples of piconets

One of the earliest proposed implementations of WPANs was known as *Bluetooth.*[8] Other technologies that were developed later for WPANs include infrared, ZigBee, and ultra- wideband (UWB). Since Bluetooth is still the most widely used technology in this space, Bluetooth architecture will be described in detail.

Bluetooth. Bluetooth (BT) is a radio frequency (RF) technology that uses the unlicensed 2.4 GHz ISM band to communicate with other Bluetooth-capable devices in the vicinity. Bluetooth uses a frequency hopping spread spectrum (FHSS) scheme to minimize interference with other nearby devices operating in the ISM band (e.g., WLAN radios) as well as nearby Bluetooth piconets. The FHSS scheme used by Bluetooth divides the ISM band into 79 frequencies. Each Bluetooth device uses a particular method of hopping between these frequencies. The hopping pattern is controlled by a single master device in a Bluetooth piconet, which can control up to seven simultaneous slave devices.

Bluetooth devices can discover each other and authenticate each other. There is provision for encrypting data exchanged between Bluetooth devices. The protocol supports both asynchronous connectionless (ACL) transfer for packet data as well as synchronous connection-oriented (SCO) transfer for voice. The initial Bluetooth specification allowed data rates up to 1 Mbps, which have been further increased to 2 and 3 Mbps in the new enhanced data rate (EDR) version. The Bluetooth Special Interest Group (SIG) certifies Bluetooth devices for compatibility with the Bluetooth specifications involving lower layers (physical and media access control or MAC) as well as the application framework. Bluetooth profiles are also defined for common usage scenarios such as headsets, cordless telephones, etc., that describe the behavior of devices for these scenarios.

Infrared. Another popular technology for short-range, low-power wireless communication is infrared (IR). It is the most commonly used technology for remote control of many consumer electronics devices. IR has also been very popular in connecting peripheral devices such as printers to personal computers. One distinguishing characteristic of IR communication is that unlike radio waves, an IR signal cannot penetrate walls. This allows an IR signal to be confined to a room, thus allowing IR devices to operate without interference from other IR devices in nearby rooms. IR components are small in size, low power, and low cost, and are therefore ideally suited for consumer appliances.

ZigBee. ZigBee is a low-cost RF technology that is designed for extremely low-power and typically low-data-rate environments such as sensor networks and industrial control applications. One major requirement of such applications is long battery life. In fact, some sensor applications rely on alkaline batteries to last up to one to two years. The battery life effectively determines the life of the sensor node itself in many applications. The reason is that sensor nodes are typically deployed in remote and often hard-to-reach locations where recharging or replacing batteries is not an option. ZigBee technology has been designed with these requirements in mind.

Ultra-wideband (UWB). UWB is an RF technology that can provide high data rates in excess of 100 Mbps for WPANs without causing interference to other wireless technologies being used in the vicinity. This is achieved by spreading a very short radio signal over a wide spectrum of radio frequencies and keeping the transmitted power low. Because of the low power and wide spectrum of frequencies used (3.1 to 10.6 GHz in the United States), UWB transmissions appear as *background noise* to other RF technologies. Promise of high data rates at short distances makes UWB ideally suited for cable replacement in multimedia applications such as digital home and consumer electronics devices because it can support video and audio streaming. The narrow pulses (1 nanosecond or less duration) used by UWB also allow its use as a technology for precise identification of location.

Relevant standards

One of the earliest standards activities in the WPAN space was the formation of an ad hoc group in IEEE Portable Application Standards Committee (PASC) in 1997 to study WPANs. On the industry side, Bluetooth SIG was formed in 1998 by five companies. These standards organizations are detailed in the following, as well as others such as the Infrared Data Association (IrDA) and the ZigBee Alliance.

Bluetooth SIG. The Bluetooth Specification 1.0 was released in 1999, and the first products for mobile phones and PCs appeared in 2000. The BT Specification 1.2 was released in 2003, followed by version 2.0 with enhanced data rates (EDR) in 2004. In 2006, Bluetooth SIG announced that it would use UWB technology from WiMedia Alliance for higher data rate Bluetooth products.[9] Today, Bluetooth SIG has more than 9,000 member companies and more than 1 billion installed Bluetooth products. Bluetooth products include mobile phones, PCs, laptops, headsets, printers, hands-free kits, MP3 players, digital cameras, televisions, watches, picture frames, alarm clock radio, sunglasses, and GPS receivers.[10]

Infrared Data Association (IrDA). The IrDA was formed in 1994 as an industry association to standardize products for infrared data transfer.[11] The IrDA-1.0 specification supports data rates of up to 115 kbps, whereas the IrDA-1.1 specification supports data rates up to 4 Mbps. IrDA devices work as point-to-point connections when they are placed within 1 m, with their transceivers facing each other. IrDA capability is found in most desktop and laptop PCs and also many digital cameras, mobile phones, and printers.

WiMedia Alliance. The WiMedia Alliance is an industry organization that creates specifications for UWB-based WPANs and certifies devices for interoperability with these specifications.[12] WiMedia Alliance UWB specifications have been chosen by Bluetooth SIG as the radio for next generation Bluetooth standards. Also, the Universal Serial Bus

Implementors Forum (USB-IF) has chosen WiMedia Alliance's UWB specification for implementation of wireless USB.

ZigBee Alliance. The ZigBee Alliance is an industry association working on specification and certification of low-power and low-cost wireless monitoring and control solutions.[13] ZigBee radios are quite popular in sensor network applications where battery-operated sensor nodes are expected to work for a year or more without replacement of batteries.

IEEE 802.15. A working group known as 802.15 was formed in 1999 within IEEE to standardize wireless personal area networks for short distances.[14] This working group later evolved into four task groups (TGs) as shown in table 2–1. TG 802.15.1 adopted a WPAN standard based on the Bluetooth 1.1 specification in 2002. The main difference between the Bluetooth 1.1 specification and the IEEE 802.15.1 standard is that Bluetooth specifies the operation of lower layers of the protocol as well as a framework for applications. In addition, Bluetooth SIG also certifies BT products for interoperability. IEEE 802.15.1, on the other hand, is restricted to the specification of lower layers of the Bluetooth technology. These are logical link control and adaptation layer protocol (L2CAP), link manager protocol (LMP), baseband, and radio. Another task group known as 802.15.2 published recommendations for coexistence of 802.15 WPANs and 802.11 WLANs in 2003. This was necessitated because many implementations of WPANs (e.g., Bluetooth) and WLANs (802.11b/g) use the same 2.4 GHz unlicensed band for communication. TG 802.15.3 was tasked to develop high-data-rate WPAN solutions.

Table 2–1. Task groups of IEEE 802.15 WPAN working group

WPAN Task Group	Description
802.15.1	Derived 802.15.1 specification from Bluetooth v1.1 specification. Published standard in 2002.
802.15.2	Published recommendations in 2003 for co-existence of Bluetooth and 802.11 based WLAN systems operating in the 2.4 GHz ISM band.
802.15.3	High data rate WPAN (11-55 Mbps) operating in the 2.4 GHz band. Draft standard including PHY and MAC layers published in 2003. 802.15.3a group was formed to explore an alternative PHY for > 100 Mbps data rates but worked was stopped due to a technical gridlock between 2 competing UWB proposals.
802.15.4	Formed to explore low data rate, long battery life WPAN solutions for sensor applications. Published a standard in 2006 and later an amendment in 2007 that added precision location capabilities.

Cost considerations

Almost all WPAN solutions are designed for small form factor consumer devices, and therefore, the cost of the WPAN part has to be small relative to the device cost. The IrDA component cost is typically less than $1, whereas Bluetooth chips cost less than $5. Initially, UWB components are likely to be expensive at approximately $20, but these costs will likely fall to the level of Bluetooth once UWB becomes popular.

Typical applications

The main characteristics of a WPAN are: short range, ad hoc networking, and low power consumption. The data rates can vary from a few kilobits per second to hundreds of megabits per second depending on the technology used. Considering all of these factors, WPANs are suitable for a whole range of applications, some of which are listed in the following:

1. Cable replacement for connections between personal devices and their peripherals or accessories. This includes low-data-rate peripherals such as headphones, as well as high-data-rate peripherals such as televisions.
2. Ad hoc networking between devices for file transfer and data synchronization.

3. Data access points, e.g., a PC connecting over a WPAN link to a mobile phone being used as a modem to access the Internet.
4. Sensor networks and industrial control applications.
5. Location identification.

Wireless Local Area Networks (WLANs)

Wireless LANs were developed primarily so that portable computers could access local area networks (LANs) such as the Ethernet without using wires. The design goals for WLANs were primarily to achieve high data rates comparable to wired LANs and enough range to cover a floor in a building. Use of unlicensed bands at 900 MHz and 2.4 GHz also provided freedom from licensing requirements. Although the first WLAN equipment appeared as NCR/AT&T WaveLAN in 1990 around the time when laptop computers were becoming widely available, using WLANs was far from easy. Early 900 MHz WaveLAN cards were available in ISA and Personal Computer Memory Card International Association (PCMCIA) form factors for desktops and laptops, respectively. However, these had to be connected to external antennas that were attached to the wall in the case of desktops or on the lid in the case of laptops. Later, 802.11-based WLAN products were introduced with smaller antennas that were part of the PCMCIA cards operating in the 2.4 GHz band. However, it was not until 2003 that the use of laptops with integrated WLAN took off, largely due to the Centrino marketing campaign by Intel.[15] Today, WLAN has found use in a variety of applications ranging from indoor LANs to hotspots in airports and cafes to outdoor metro Wi-Fi networks.

Architecture and components

WLAN capable devices can typically operate in two modes: *infrastructure* and *peer-to-peer.* These two modes are described next.

Infrastructure WLAN. An infrastructure WLAN (fig. 2–3) is controlled by a special device known as the *access point (AP),* which provides network access to WLAN client stations (STAs). Desired WLAN coverage area is usually served by a network of APs—their number determined by a tradeoff between the capacity required and the mutual interference that can be tolerated. APs can themselves be connected together by a wireless distribution system (WDS) or they may be connected to the wired LAN (Ethernet). Each wireless LAN or service set (SS) uses a unique service set identifier (SSID) that is periodically broadcast by each AP that belongs to the same WLAN. In addition, each AP and the clients connected to the WLAN via the AP are known as a *basic service set (BSS).*

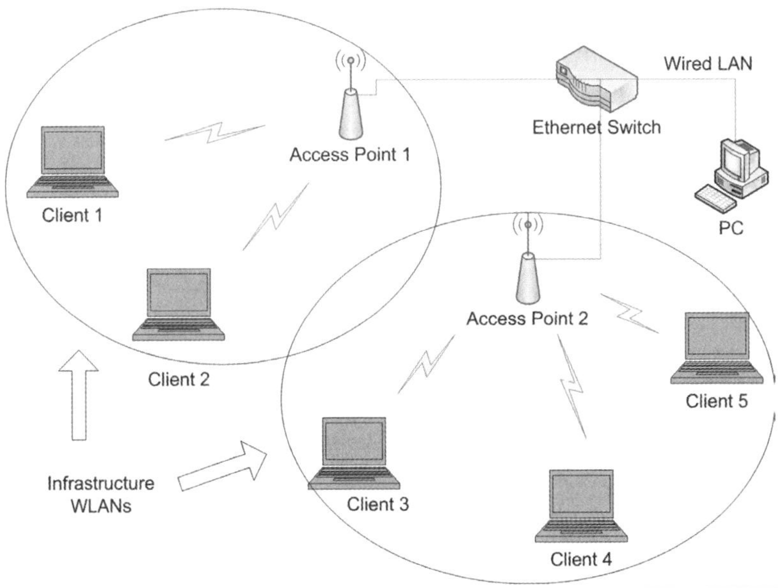

Fig. 2–3. Example of infrastructure WLAN

In an infrastructure network, most of the traffic flows from WLAN clients to the wired network and the Internet via the AP. If two clients who belong to the same BSS wish to communicate with each other, they can either do so via the AP or set up a direct link with each other with the AP's cooperation. The basic protocol used in a BSS for multiple access is carrier sense multiple access with collision avoidance (CSMA/CA). This is a variant of the carrier sense multiple access with collision detection (CSMA/CD) protocol user over wired Ethernet. A WLAN radio can either transmit or receive at a given time but typically cannot do both, and therefore collision detection is not possible. Hence CSMA/CA is used. Most WLANs used in enterprise and hotspot deployment are infrastructure WLANs.

Peer-to-peer WLAN. A peer-to-peer (or infrastructureless) network (fig. 2–4) involves communication between two WLAN clients (STAs) without any AP. This mode is typically used for ad hoc networking, e.g., between two laptop computers. Using the peer-to-peer mode, WLAN devices can even form a mesh network consisting of multiple point-to-point links. In a mesh network, data from a WLAN device may travel over multiple wireless hops before it reaches its destination within the mesh network or on a wired network if the wireless mesh network is connected to one. Each WLAN device in a mesh network can forward data from its neighbors to other neighbors, thus performing a *routing* function similar to Internet routers. Such mesh networks can be deployed relatively quickly in areas with no Internet connectivity or in disaster situations where phone and data networks may be down because of an earthquake or storms.

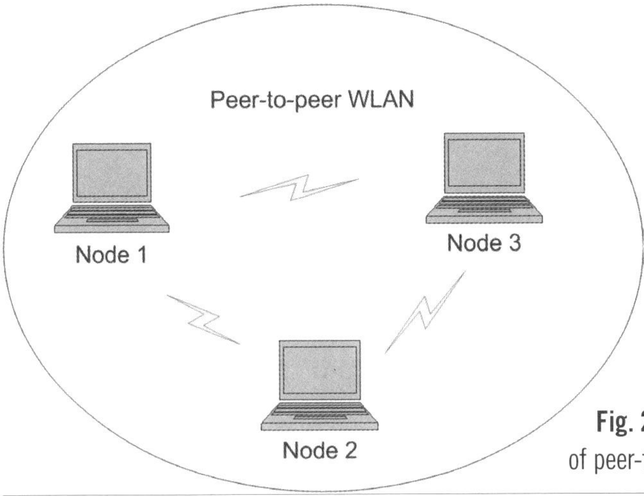

Fig. 2–4. Example of peer-to-peer WLAN

Relevant standards

IEEE 802.11 Working Group. The most commonly used WLAN standard is published by the IEEE 802.11 Working Group.[16] The 802.11 group was established in 1990 and the original 802.11 WLAN standard was published in 1997. The base 802.11 standard provided for three physical (PHY) layers: infrared, 2.4 GHz frequency hopping spread spectrum (FHSS), and direct sequence spread spectrum (DSSS). The maximum data rate possible was 2 Mbps. The 2.4 GHz PHY layers were designed to operate in the 2.4 GHz industrial, scientific, and medical (ISM) band, where operation of low-power radio devices is allowed without a license.

Two major revisions to the 802.11-1997 standard were published in 1999. The 802.11a provided a PHY layer based on orthogonal frequency division multiplexing (OFDM) using the 5 GHz unlicensed national information infrastructure (UNII) band with speeds up to 54 Mbps. The 802.11b standard provided a higher data rate PHY layer to operate at 11 Mbps in the 2.4 GHz band. Later, many amendments to the 802.11 standard were published with different letters used to denote the different task groups that worked on these amendments. Some of these amendments are given in the following:

- **802.11e.** Added quality of service (QoS) features to 802.11 MAC layer.
- **802.11g.** Added a higher data rate version of PHY layer (up to 54 Mbps) based on OFDM operating in the 2.4 GHz ISM band that is backward compatible with 802.11b PHY.
- **802.11i.** Added security and authentication mechanisms to 802.11 MAC layer.

There are other task groups currently active under the IEEE 802.11 working group that are expected to complete additional amendments to the 802.11 standard. Some important ones are given in the following:

- **802.11n.** Proposal includes changes to PHY and MAC layers to increase data rates to 200 Mbps or more.
- **802.11s.** Provides for multihop wireless mesh networks that can configure themselves automatically.

Wi-Fi Alliance. The Wi-Fi Alliance is an industry organization that was formed in 1999 to facilitate interoperation of wireless LAN devices based on the IEEE 802.11 standard.[17] (*Wi-Fi* is short for "wireless fidelity.") Interoperability is achieved by publishing interoperability specifications for each category of devices (e.g., client station or access point) and supported functionality (e.g., 802.11a/b/g, network security, or quality of service). Wireless LAN products are then certified for interoperation in designated laboratories.

Cost considerations

WLAN products have become extremely popular over the years since the first NCR/AT&T WaveLAN product was introduced in 1990. Some reasons for this popularity are its operation in the unlicensed ISM band, ability to work with or without infrastructure support, and ease of use. Because of this popularity, WLAN chipsets costs have steadily come down with mass production. Today, WLAN client solutions in embedded form factors (e.g., integrated with laptops, mobile phones, and PDAs) cost only a few dollars, while basic WLAN access points can be purchased for as little as $20.

Typical applications

Along with the increasing popularity of WLAN devices in traditional networking applications, new usage models for such devices continue to evolve as price points for WLAN chipsets continue to fall. Here is a nonexhaustive list of applications in which WLAN devices have been used so far:

- **Wireless extension of LAN.** This was the primary application for WLANs.
- **Ad hoc networking.** For example, peer-to-peer communication between two neighboring WLAN devices.
- **Hotspots.** For providing Internet access to portable computers at airports, hotels, convention centers, cafes, train stations, etc.
- **Outdoor wireless coverage.** For example, metro Wi-Fi solutions to provide coverage over an area within a city with Internet access.
- **Mesh networks.** These could be multihop ad hoc networks where each WLAN device can also forward network packets from other nearby WLAN devices in addition to sending and receiving its own network packets. Alternatively, these could be planned multihop networks that can carry network packets over long distances (up to few miles and even tens of miles) using

point-to-point wireless links enabled by highly directional antennas.[18]

- **Wireless sensor networks.** This application of WLAN has become possible thanks to the shrinking size of WLAN chipsets and also because of their dropping price points. By lowering the *duty cycle* (the amount of time that a device is active versus the amount of time it is idle) of a WLAN device, it is possible to operate WLAN-based sensor nodes on alkaline batteries that can last as long as a few years.
- **Location identification.** Using multiple access points that accurately know their own coordinates and triangulation methods, the location of a WLAN device can be determined to an accuracy of a few meters. This is quite useful in indoor environments where other location identification techniques such as GPS do not work well.
- **Unlicensed mobile access (UMA).** Dual-mode mobile phones that can work with both cellular and Wi-Fi networks have been recently introduced for such an application. A subscriber with such a phone can make or receive calls using a Wi-Fi network if one is available (e.g., at home or in a hotspot), with the phone call automatically switching to the cellular network if the subscriber walks out of the Wi-Fi coverage area. A mobile phone subscriber can thus save money on calls made over Wi-Fi networks, while the network operator benefits by off loading users in a Wi-Fi coverage area from its cellular network, thus saving the capacity for users who are truly mobile.[19]

Wireless Metropolitan Area Networks (WMANs)

WMANs extend wireless network connectivity to cover entire cities or metropolitan areas. Although a single WMAN base station could potentially cover a metro area with high-speed wireless Internet access, capacity considerations typically require using a cellular model where each base station covers an area of a few square kilometers. WMANs can be used to provide fixed broadband wireless access (BWA), also known as *last mile access*, similar to cable Internet and DSL. WMANs can also be used to provide mobile Internet access at pedestrian and vehicular speeds to portable computers and PDAs.

Although many network technologies including local multipoint distribution service (LMDS) and multichannel multipoint distribution service (MMDS) can be classified as WMANs, an emerging technology known popularly as WiMAX (Worldwide Interoperability for Microwave Access) is likely to be deployed as the predominant WMAN technology globally for residential and commercial Internet access as well as mobile Internet access. WiMAX comes in two flavors: licensed and unlicensed, depending on the frequency spectrum used. Licensed WiMAX solutions typically operate in two major frequency bands of 2.3–2.5 GHz and 3.3–3.8 GHz. However, recent developments indicate that another band in 700 MHz will also be available for WiMAX in some countries, including the United States. Unlicensed WiMAX solutions typically operate in the 5.8-GHz band, although it is possible to use WiMAX also in the 2.4 GHz unlicensed band subject to local and country-specific regulations.

Architecture and components

Figure 2–5 shows the architecture of a WiMAX network. Each base station (BS) covers a certain area around it for network access and may typically serve a mix of fixed and mobile clients within its coverage area. Base stations in a geographic area are connected via a backhaul to access service network gateway (ASN-GW). ASN-GW is connected via traditional IP network links to a core services network (CSN). The CSN includes authentication, authorization, and accounting (AAA) services as well as home agent (HA) services for operation of mobile IP (MIP).[20] ASN-GW, on the other hand, provides the foreign agent (FA) service of mobile IP that allows roaming mobile terminals to connect to an ASN of a different service provider than their own. Interfaces are defined for the following interaction within a WiMAX network: mobile subscriber station (MSS) to BS, BS to ASN, ASN to another ASN, ASN to CSN, and CSN to another CSN. All these interfaces operate on IP-based protocols with support for both IPv4 and IPv6.

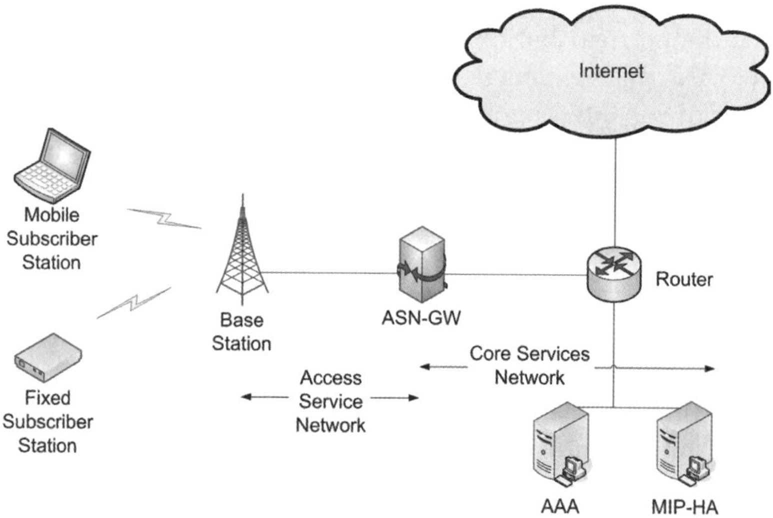

Fig. 2–5. WMAN network architecture

Relevant standards

IEEE 802.16 Working Group. This is the primary standardization forum for WMANs.[21] This working group was established in 1999, and the first 802.16 standard was published in 2001. The original 802.16 standard was designed for operation in *line of sight (LoS)* mode using licensed spectrum in the 10–66 GHz range. The applications envisaged were mainly broadband wireless access and backhaul for the enterprise. Subsequently, another variation of the original standard, namely 802.16a, was published in 2003 for non–line of sight (NLoS) operation in the 2–11 GHz licensed as well as unlicensed spectrum. In 2004, another specification known as 802.16-2004 or 802.16 Rev D was ratified by IEEE, which superseded all the earlier 802.16 standards.[22] This new standard was designed for fixed broadband wireless access (BWA) and included LoS operation in the 10–66 GHz licensed spectrum as well as NLoS operation in licensed and unlicensed spectrum below 11 GHz. The 802.16-2004 standard also harmonized a potentially competing European standard known as European Telecommunications Standards Institute (ETSI) HiperMAN standard within its specification. In 2005, an amendment to the 802.16 standard, known as 802.16e or 802.16 Rev E, was ratified with support for client mobility at pedestrian and vehicular speeds.[23] Two task groups within the 802.16 working group are currently working on extensions to 802.16e to support mobile multihop relays (TG 802.16j) and advanced air interface (TG 802.16 m) that will enable higher data rates.

WiMAX Forum. The WiMAX Forum is an industry organization similar to the Wi-Fi Alliance, and it certifies WiMAX equipment based on IEEE 802.16 standards for interoperability and compliance. The WiMAX Forum has many working groups that work on interoperability specifications for various layers in a WiMAX network, e.g., application, networking, roaming, and air interface layers. Many products have been certified in the base station and the subscriber station categories based on the older 802.16-2004 (fixed WiMAX) standard. Recently, in June 2008, The WiMAX Forum certified several base station and mobile station module products based on the 802.16e (Mobile WiMAX) standard operating in the 2.5 GHz band. More such certifications are expected before the end of 2008. The WiMAX Forum also works on promoting WiMAX standards worldwide and works with the regulatory bodies in various countries on spectrum allocation for WiMAX operation.

Cost considerations

WiMAX equipment is still relatively expensive compared to WLAN equipment because WiMAX standards are fairly new. Typical subscriber stations based on the 802.16-2004 standard are available for a few hundred dollars, whereas base stations typically cost anywhere from $10,000 to $100,000. Most WiMAX products available today operate in one of the licensed bands (2.3 to 2.5 GHz and 3.3 to 3.8 GHz). Products based on 802.16e standard have started to appear in the market but interoperability between vendors will be a concern until these products are certified by the WiMAX Forum. Some original equipment manufacturers (OEMs) have also started shipping WiMAX subscriber cards for laptops in the PCMCIA form factor. However, very few deployments of 802.16e-based WiMAX networks exist in the world today where these can be used.

The 802.16e subscriber station prices (especially for portable form factors) are expected to quickly drop to about $100 or less over the next year or two. However, base station costs are expected to remain high ($100,000 or more) in the near term. WiMAX networks are expected to be deployed by service providers using licensed spectrum at least in urban and suburban areas. Large infrastructure investments are needed to cover metropolitan areas, involving communication towers, backhaul networks, power infrastructure, radio assemblies, and antennas. Use of such carrier networks will typically involve subscription-based service similar to Internet access and mobile phone subscription plans.

Typical applications

There are many applications for which WiMAX networks are likely to be used:

- **Fixed broadband wireless access.** Last mile wireless solution for Internet access to residential or enterprise customers similar to DSL, cable, and leased lines.
- **Mobile broadband Internet access.** It is expected that WiMAX networks based on 802.16e will support broadband data rates up to a few Mbps for mobile data networking across cities and towns and also in trains and on highways. Many applications such as Internet browsing, audio and video streaming, and even video conferencing should be possible in such a scenario.
- **Mobile voice over IP.** Since WiMAX provides an IP-based network infrastructure, it can be used for voice calls in a way similar to mobile phones. However, since the WiMAX network is not as optimized for voice calls and handoffs as cellular telephony networks, performance of voice calls over WiMAX may not be as good as mobile telephone calls.
- **Rural connectivity.** WiMAX is the most promising technology for providing broadband connectivity over long distances, which is typical of rural areas, especially where wired broadband networks such as fiber do not exist.
- **Wireless backhaul.** WiMAX can be used to provide backhaul in place of fiber or DSL for connecting Wi-Fi hotspots or multiple campuses of an enterprise or university.

Wireless Wide Area Networks (WWANs)

The term *WWAN* is typically used for mobile (cellular) telephone networks that support some form of data networking in addition to voice calls. Cellular telephone network implementers have spent a lot of effort in making cellular coverage as ubiquitous and seamless for mobile users as possible. A mobile user making a cellular phone call can hardly notice when the call is handed over from one cellular base station to another. As laptops and PDAs started becoming popular in the late 1990s, the need for accessing the Internet while on the move also started growing. One of the earliest cellular data standards known as cellular digital packet data (CDPD) was implemented as an overlay network over the first generation analog advanced mobile phone service (AMPS) cellular networks around 1995. The speeds offered were comparable to dial-up modems, which was a popular method for accessing the Internet from home at that time.

Second generation (2G) cellular networks based on CDMA, TDMA, and GSM standards also had provision for data networking with speeds ranging from 9.6 to 14.4 kbps. The main application for cellular data on 2G cellular networks was mobile fax. A cellular user could also use 2G technologies to connect to the Internet using the cellular connection as a dial-up line. Since data was carried in these networks on one or more voice channels that required a connection between the mobile terminal and the cellular network, this form of data access was termed *circuit-switched data*.

Third generation (3G) cellular networks are based on the International Mobile Telecommunications-2000 (IMT-2000) vision of the International Telecommunication Union (ITU). The IMT-2000 standard aims to provide a global telecommunication standard for cellular voice and IP-based packet data that utilizes uniform spectrum allocation around the world and defines standard interfaces for operations such as billing and roaming across network components. The 3G network specifications provide for data rates ranging from 144 kbps at vehicular speeds to 2 Mbps for stationary subscribers. Before 3G networking equipment and handsets were available, many evolutionary schemes were developed to provide higher data rates than the 2G limit of 14.4 kbps. These schemes were collectively known as 2.5G and include specifications such as general packet radio service (GPRS), enhanced data rates for GSM evolution (EDGE), and CDMA2000 phase 1.

Architecture and components

Fig. 2–6 shows the architecture of a 3G/UMTS network. On the left side of the figure is the UMTS terrestrial radio access network (UTRAN), which consists of the node B (equivalent of a base station in GSM), the radio network controller (RNC), and user equipment (UE) shown in the form of a mobile phone and a PDA. Each RNC can control multiple node Bs. The RNC can also handle traditional GSM/GPRS radio access networks (RANs). RNC separates data packets and voice calls. Data packets are sent via the serving GPRS support node (SGSN) and gateway GPRS support node (GGSN) towards the Internet, while voice calls are sent via the mobile switching center (MSC) towards the PSTN/ISDN telephone network.

Gateway MSC (GMSC) connects the UMTS network to the PSTN/ISDN side. Media gateway (MGW) converts packetized voice into circuit-switched voice for PSTN/ISDN network and vice versa. Each MSC also serves as a visitor location register (VLR) in the traditional cellular network. The home location register (HLR) maintains a mobile user's subscription information as in a traditional cellular network and is consulted by MSCs as well as SGSN and GGSNs for a user's profile.

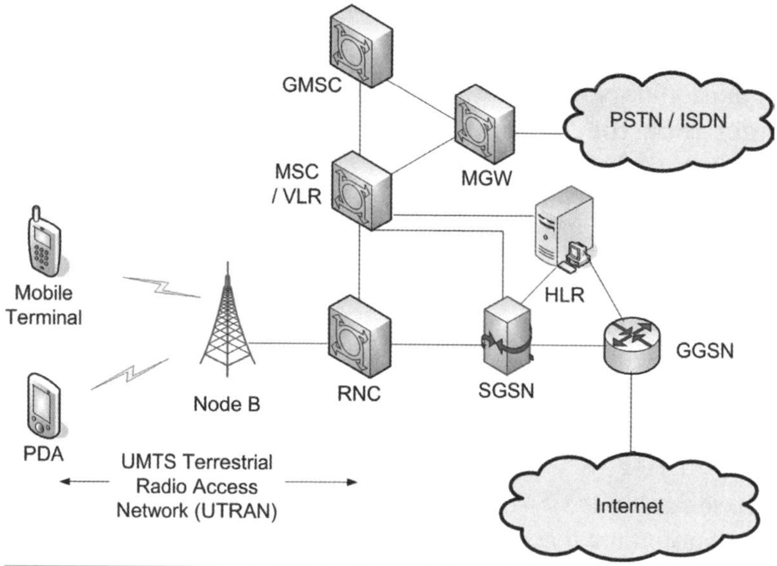

Fig. 2–6. Third generation (3G) UMTS network architecture

Relevant standards

ITU. The International Telecommunication Union (ITU) is a United Nations (UN) agency responsible for standardization of communication technologies.[24] The ITU started working on the "International Mobile Telecommunications (IMT)" vision in the mid-1980s. The IMT-2000 technical specifications were adopted in 2000 as a globally interoperable set of standards for the third generation of mobile telephony. The vision includes a variety of services such as voice, data, and multimedia applications running over a common wireless network. The two main terrestrial (i.e., nonsatellite) radio technologies included in IMT-2000 are both based on wideband CDMA. These are W-CDMA and CDMA2000. W-CDMA is designed for an upgrade from GSM and GPRS networks, while CDMA2000 is designed for an upgrade from CDMA-based 2G cellular networks. ITU has also recently approved the inclusion of

Internet protocol—orthogonal frequency multiple access (IP-OFDMA) or WiMAX as another terrestrial radio technology into IMT-2000.

ETSI. The European Telecommunications Standards Institute (ETSI) is an independent organization working on telecommunications standards for Europe that can also be used worldwide.[25] ETSI published the Global System for Mobile Communications (GSM) standard for cellular telephony in 1990. Later, a packet data service called *general packet radio service (GPRS)* was added to GSM in 1997 to allow data rates up to 100 kbps. In 1999, another standard known as *Enhanced Data Rates for GSM Evolution (EDGE)* was published that increased the data rates up to 384 kbps.

3GPP. The 3rd Generation Partnership Project (3GPP) group was created in 1998 to develop a 3G mobile communications standard for GSM networks.[26] It includes ETSI as one of the members. The 3GPP specification for 3G networks is known as *W-CDMA* or the *Universal Mobile Telecommunications Standard (UMTS)*, which is part of IMT-2000. The 3GPP has also published enhanced standards that provide higher data rates than in UMTS. High Speed Downlink Packet Access (HSDPA) allows downlink data rates of up to 14.4 Mbps, but the uplink rates are limited to 384 kbps. High Speed Uplink Packet Access (HSUPA) allows uplink rates up to 5.76 Mbps. Currently, 3GPP is working on a fourth generation (4G) mobile communications standard named *Long-Term Evolution (LTE)*.

3GPP2. The 3rd Generation Partnership Project 2 (3GPP2) is another group that has published 3G standards collectively known as the CDMA2000 family for evolution of 2G CDMA–based networks.[27] CDMA2000 1xRTT is a 2.5G technology developed by 3GPP2 with effective data rates up to 144 kbps. A newer CDMA2000 1xEVDO (Evolution Data Optimized) Rev. A standard allows data rates up to 3.1Mbps.

WAP. Wireless application protocol (WAP) was initially developed by WAP Forum, which is now part of Open Mobile Alliance.[28] WAP is designed to accelerate Web-browsing performance over 2G and 2.5G cellular networks and is very popular for accessing the Internet from smart phones and PDAs. WAP requires carriers to deploy additional equipment in their networks, which essentially strips off graphics from a Web page before serving it to a mobile user. Popular Web sites often create their WAP versions for mobile users, thus ensuring faster download of their Web pages on small devices connected via limited bandwidth networks such as GSM, GPRS, and EDGE.

Cost considerations

WWANs have been deployed in one form or the other in most countries of the world. Most of these deployments are 2G networks based on GSM or CDMA. The cost of GSM and CDMA handsets have been falling steadily for the last few years and some of the models designed for emerging markets can be purchased for as little as $25. Along with an increase in the number of cellular subscribers worldwide, usage costs have fallen to as little as 1¢/minute in some developing countries. Unlike the United States, where most cellular callers are charged for airtime used for incoming as well as outgoing calls, in most other countries, it is always the calling party that pays for a call. This charging method along with the popularity of prepaid cellular connections have given rise to new usage models where people who cannot afford paying for cell phone calls use mobile phones mostly to receive calls. Typical data access plans for 2.5G networks, however, cost anywhere from $10 to $100 per month for a few gigabytes of data transfer. Handsets and PDAs that can use 2.5G networks for data access are typically priced higher than entry level phones. Prices of 3G handsets are still comparatively high at a few hundred dollars, but these are expected to fall as 3G networks are deployed in developed countries and the economies of scale start to kick in.

Typical applications

There are a number of applications that have been used in the past and are likely to be used in the future over WWANs:

- Voice calls (circuit switched as well as packetized voice and voice over IP)
- Narrow-band data access, including using a cell phone as a dial-up modem
- E-mail access and synchronization with office computers, e.g., with BlackBerry and similar devices
- Internet access, e.g., Web browsing using WAP
- Wireless fax
- Remote access to business applications such as sales automation, field sales support, and accessing corporate databases
- Operation of point-of-sale terminals and ATMs where no wired connectivity is available
- Broadcast of audio/video content
- Text messaging via Short Message Service (SMS), Multimedia Messaging Service (MMS), and sharing of pictures and video clips
- Downloading music, TV shows, ringtones, podcasts, and even movie clips
- Location tracking, including E-911 service that allows location identification of a mobile device within a few hundred feet

Summary

In this chapter, various wireless networking technologies in use around the world today have been described, along with their applications. Table 2–2 below gives a comparative summary of these technologies using various parameters such as range, cost, frequency spectrum, and applications.

Ajay Bakre is the principal author for chapter 2 on wireless networking technologies.

Table 2–2. Brief comparison of wireless networking technologies

Wireless Network	Typical Range	Data Rates	Typical Cost	Frequency Spectrum	Typical Applications
WPAN	< 10m	1-3 Mbps (BT) / 480 Mbps (UWB)	$5-25 per component	2.4GHz (BT) / 3.1-10.6 GHz (UWB)	Short range, cable replacement, ad hoc / sensor networks
WLAN	< 100m	1 Mbps (802.11) to 248 Mbps (802.11n)	$20-$50 for client / access point	2.4GHz (11b/g/n) / 5GHz (11a/n)	Indoor Wireless connectivity for portable computers
WMAN	< 10 km	1-5 Mbps per user	$200-300 (CPE) / $10-50K (BS)	700MHz, 2.3-2.5GHz, 3.3-3.8GHz, 5.8GHz	Fixed / mobile broadband data access
WWAN	< 30 km	9.6 kbps (2G) to 2 Mbps (3G)	$25-$200 (MT) / $10-50K (BS)	800-900MHz, 1.8-2.2GHz	Wide area connectivity for mobile phones / PDAs

References

1. Abramson, N. 1985. Development of the ALOHANET. *IEEE Transaction on Information Theory.* Vol. IT-32, no. 2. March.

2. Wikipedia. ALOHAnet. http://en.widipedia.org/wiki/ALOHAnet.

3. Smith, C., and D. Collins. 2002. *3G Wireless Networks.* Boston, MA: McGraw-Hill.

4. History of Bluetooth technology. http://www.bluetooth.com/Bluetooth/SIG/History_of_the_SIG.htm.

5. ZigBee Alliance. http://www.zigbee.org.

6. WiMAX Forum. http://www.wimaxforum.org.

7. Salmi, J. 2006. WPAN—wireless personal area networks. Helsinki University of Technology, SMARAD Centre of Excellence. February 21. http://www.comlab.hut.fi/opetus/4210/presentations/25_wpan.pdf.

8. Bluetooth Special Interest Group. http://www.bluetooth.com.

9. Bluetooth SIG press release. 2006. Bluetooth SIG selects WiMedia Alliance ultra-wideband technology for high speed Bluetooth applications. March 28. http://www.bluetooth.com/Bluetooth/Press/SIG/BLUETOOTH_SIG_SELECTS_WIMEDIA_ALLIANCE_ULTRAWIDEBAND_TECHNOLOGY_FOR_HIGH_SPEED_BLUETOOTH_APPLICATION.htm.

10. History of Bluetooth technology. http://www.bluetooth.com/Bluetooth/SIG/History_of_the_SIG.htm.

11. Infrared Data Association (IrDA). http://www.irda.org.

12. WiMedia Alliance. http://www.wimedia.org.

13. ZigBee Alliance. http://www.zigbee.org.

14. Institute of Electrical and Electronics Engineers. IEEE 802.15 Working Group for WPAN. http://www.ieee802.org/15.

15. Wikipedia. Centrino. http://en.wikipedia.org/wiki/Centrino.

16. Institute of Electrical and Electronics Engineers. IEEE 802.11 Working Group. http://www.ieee802.org/11.

17. Wi-Fi Alliance. http://www.wi-fi.org.

18. Patra, R., et al. 2007. WiLDNet: design and implementation of high performance Wi-Fi based long distance networks. Proc. 4th USENIX Symposium on Networked Systems Design and Implementation (NSDI). April.

19. Unlicensed mobile access (UMA) technology. http://www.umatoday.com

20. Perkins, C., ed. 2002. IP mobility support for IPv4. Internet Engineering Task Force (IETF), Request for Comments: 3344. August.

21. Institute of Electrical and Electronics Engineers. IEEE 802.16 Working Group. http://www.ieee802.org/16.

22. Institute of Electrical and Electronics Engineers. 2004. IEEE Standard 802.16-2004. Air interface for fixed broadband wireless access systems. October.

23. Institute of Electrical and Electronics Engineers. 2006. IEEE Standard 802.16e-2005. Air interface for fixed and mobile broadband wireless access systems—amendment for physical and medium access control layers for combined fixed and mobile operation in licensed bands. February.

24. International Telecommunications Union. http://www.itu.int.

25. European Telecommunications Standards Institute. http://www.etsi.org.

26. 3rd Generation Partnership Project. http://www.3gpp.org

27. 3rd Generation Partnership Project 2. http://www.3gpp2.org.

28. Open Mobile Alliance. http://www.openmobilealliance.org.

3 APPLICATION OF RFID AND WIRELESS TECHNOLOGIES FOR MOBILE ASSET MANAGEMENT

Business Process Challenges Addressed by Mobile Asset Tracking

The improved visibility and tracking of assets (tools and parts, material, personnel, vehicles) is rising to the top of many CEO and board agendas in the utility industry. This capability for mobile asset tracking provides the potential of reducing costs and working capital and improving the service provided by utilities. Asset tracking in the utility industry impacts a number of business processes that could benefit from the usage of RFID technology.

Tools and parts management

The problems with tools and parts can be related to misplacement and loss. Within a warehouse, this can be because of misstocking, mispicking, or misshipping (akin to misissuance). *Misstocking* refers to the human- or machine-induced error of accidentally putting the product in wrong bins. Typically the misstocked inventory is as good as lost inventory until it is discovered and identified during the next cycle count, at a huge operational expense. This moreover leads to wrong inventory being present in wrong bins and essentially not having enough of the right products in the right bins. Misstocked inventory leads in turn to lost sales, customer frustration, and excessive inventorying by the planners and buyers as a compensatory action.

Similarly *mispicking* occurs when the wrong products are accidentally picked for satisfying a demand. This can occur in large automated warehouses.

Misshipping here refers to the situation when in response to a request for tools or parts, the wrong products are issued. This results in either the wrong quantities of the right product showing up, i.e., overages or underages, or simply the wrong products showing up as misships or misissuance. Most current operations find it hard to detect the errors at the very outset, because the shipments are usually received at a case or pallet level. The traditional bar code technology cannot penetrate inside the boxes to see what exactly is being received. Errors are caught, if at all, at the case or pallet level, and then the products are set aside for "problem-receive" that introduces a significant and costly delay in handling as well as occupying precious floor space.

Loss of tools can occur due to wear and tear, accidental misplacement, or shrinkage. It is normal to expect that a certain number of tools will be lost due to wear and tear. However, it is important that the loss of certain critical tools and parts be tracked because this can help reduce the rate of loss of tools.

Materials management/shelf-life management

Chemical products used in utilities have a finite shelf life. While fulfilling orders in a warehouse, it is necessary to implement FEFO (first expired first out) sequences with the desired effect that products with the least remaining shelf life get moved out from the inventory first. The pick sequences need to be facilitated by backend supply chain systems that maintain the shelf-life information. Current bar code technologies make it difficult to locate the items that need to be moved out first, and typically this requires a human with a bar code scanner to manually locate the items. In large warehouses this can be expensive and time-consuming. Using RFID it is possible to locate the exact item from a distance, and using conveyor systems, the pick sequence can be rapidly implemented in an automated manner.

Hazardous materials (hazmat) management

Hazardous materials are substances or combinations of substances that have the capability of harming people's health in particular or the environment in general. Many of the products or by-products needed to support a utility industry are hazardous; hence, hazardous materials are manufactured, stored, used, or transported across the operations within a utility.

A variety of shelving and racking systems is required to meet all types of hazmat FDA regulations. *Hazmat labeling* is the process of identifying, classifying, and properly labeling all hazardous materials. The hazmat labeling allows the materials to be unmistakably recognized as such by the personnel and equipment involved in handling the products, both within the warehouse and for transportation purposes outside of the warehouse. Hazmat management goes hand in hand with shelf-life management, as the shelf lives of products may be closely associated with their hazardous characteristics. Hazmat management is an essential and complex component in the utility and chemical logistics.

A significant challenge in hazmat transportation is the need to avoid areas with significant human population such as towns and cities. It is important to alert first responder personnel (police, fire, emergency) should there be a problem during hazmat transportation. RFID, with its real-time track and trace capabilities, can efficiently manage hazmat handling and transportation.

Lot control (tracking and traceability)

In addition to inventory location, chemicals and hazardous materials within a utility need to be tracked by the lot number (lot tracking), as well as by the expiration date and the manufacturer (traceability). All these operational problems end up in enterprise systems housing inventory information that does not reflect reality. This leads to significant problems that ripple throughout the enterprise. For example the inventory system thinks that there is inventory of 1,000 bottles of a product X present in the right shelves that belong to product X. As a result the warehouse takes 10 orders of 100 bottles each. The reality of the inventory is that only 850 bottles of product X are in the right shelves because of the problems mentioned in the previous section. As a result when the pick wave is issued, only 8 work orders are satisfied and the rest are delayed, partially filled, or cancelled. Moreover if this discrepancy is not caught, the next planning and buying cycles will continue to order insufficient quantities of product X, leading to dissatisfied customers and reduced efficiency in the enterprise.

Personnel tracking

Proper personnel access control is essential to the utility industry's high-security facilities. However, the need for security may cause challenges when it increasingly converges with differing needs. Security needs must be weighed against responsibilities related to the ability to respond to fire and natural disasters and to address safety concerns. During certain events, there may be an immediate need to accomplish the following:

- Provide immediate exit access to every personnel in the building
- Provide a count of how many personnel have left the building
- Provide a count of how many personnel are still in the building
- Track and locate in real time the remaining personnel in the building
- Provide ways and means of alerting the remaining personnel in the building

- Develop a fire safety system that works with the facility's fire detection system, alarm system, and fire extinguishers

Currently it is very difficult to track personnel as they are moving inside facilities and to trigger alerts if unauthorized personnel gain access to secure facilities. Further, in case of fire and emergency, it is difficult to determine the number of personnel remaining in the facility.

Fleet management

Utilities manage fleets that have large numbers and types of vehicles. These fleets are required for accessing distributed sites for conducting repairs, managing routine operations, and in some instances, hauling and shipping chemicals, waste products (such as biosolids), and other materials. To improve the management of fleets, a fleet manager would like to accomplish the following:

- Perform usage planning and optimization and financial analysis to achieve better efficiency and end-to-end delivery time performance
- Collect and report shipment status through automated means
- Measure, manage, and report performance in accordance with contractor predetermined performance criteria
- Resolve claims for loss and damage
- Possess or have access to sufficient working capital to make timely investments needed to enable continuous improvement

The business process challenges are not new challenges. Although there have been many efforts to address these issues in the past, few have achieved the promised level of success—primarily because of a lack of focus on the processes of "track, sense, and respond" and the lack of technologies that enable this capability. In the technology arena, however, new technologies have emerged that address many of the major problems. These technologies include RFID, wireless infrastructures, and distributed computing. Coupled with improved processes, these can provide significant benefits to utilities. This section describes usage of RFID for mobile asset tracking, details the benefits, and discusses how the challenges could be solved for tracking and management of assets handling.

Assembling an RFID Solution for Mobile Asset Tracking

As discussed before, tracking mobile assets is critical for the utility industry. This section will examine the steps needed to use passive RFID for mobile asset tracking. Later in the chapter, the use of active RFID and GPS tags for tracking assets will be discussed.

Site survey and RFID readiness

Geography and location of the site plays an important part in determining the RFID infrastructure that will be used. The first step is to conduct a site survey and determine the allowable frequency band for passive RFID tags. Here the concept of a layered approach for tagging at each logistics unit will be introduced. The layers of a logistics unit have been defined by the U.S. Department of Defense (DOD) and are illustrated in figure 3–1.

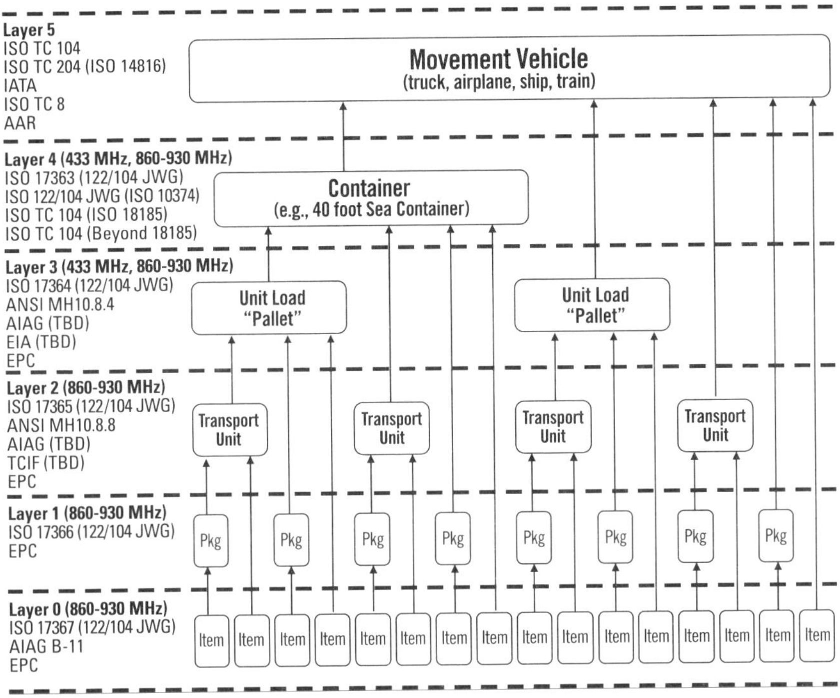

Fig. 3–1. Layers of a logistics unit

The next step is to apply tags at the appropriate layer of the logistics unit. Inside the depot or a distribution center, the usual layers are the pallet layer, the case layer, and the item layer. Figures 3–2 and 3–3 demonstrate how one would apply tags at each of these layers. Great care is needed while applying tags to metals and liquids as there are readability issues with them due to scattering or absorption of RF waves. Choosing the right labeling material for the passive tag is the key to solving this problem. Sometimes, in applications involving toolings or metal casings, the bolt tag has demonstrated maximum readability.

Fig. 3–2. Applying the bolt tag for tracking metal equipment

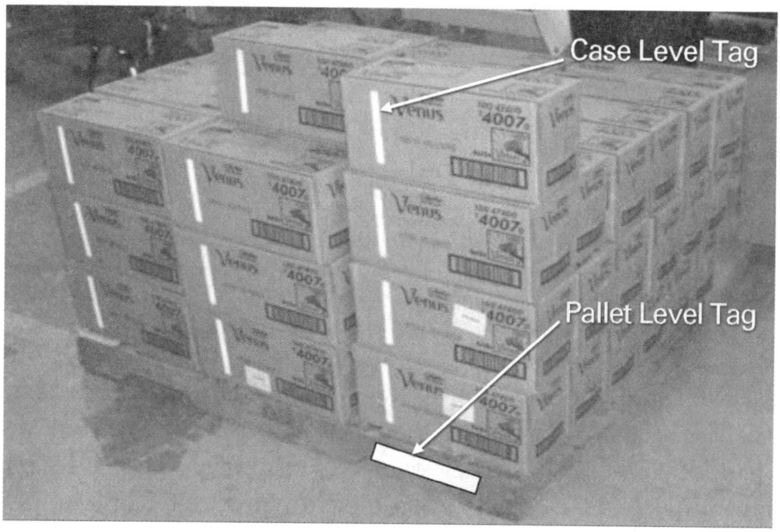

Fig. 3–3. Tags applied at the case and pallet level

The next focus is on instrumenting the portals or the dock doors for RFID. The main equipments for this task are the RFID readers, RFID antennas, RFID handhelds, and the networking infrastructure. For maximum readability, there is usually more than one antenna attached to the reader. The reader is networked to the site IT infrastructure via a LAN or a wireless LAN. It is becoming more commonplace to plug RFID devices to a WLAN in order to allow for hands-free operation. RFID readers are attached to the receiving dock to do operations such as receiving, reconciling, and inventory, while on the shipping side, they enable the generation of serialized advance ship notices. By applying suitable heuristics to items coming and leaving the site, it is possible to get an accurate count of item-level information at the site.

Increasingly the RFID infrastructure is tied to other physical devices such as PLCs for controlling the conveyor systems, light sensors, and infrared scanners to automate the flow of materials. Figure 3–4 illustrates the RFID instrumentation of dock doors.

Fig. 3–4. Receiving dock door instrumented with RFID reader, antennas, and portal

Using EPCglobal Network for RFID data management

RFID is a key enabler in bringing real-time visibility, security, and efficiency for mobile asset tracking. However, pervasive use of RFID tags for mobile asset tracking within the enterprise is bound to generate an enormous amount of data. Further, tying all the physical assets of the enterprise with the virtual systems requires new capabilities and redesign of IT infrastructure within the enterprise. The RFID standards organization—EPCglobal—has done significant work in minimizing RFID impact on existing IT infrastructures. This section will examine some of the key software standards and protocols ratified by EPCglobal that will enable real-time mobile asset tracking.

EPCglobal Network software architecture components

The EPCglobal Network architecture (fig. 3–5) shows the high-level components of the EPC Network. This discussion will focus on only the key software components as they apply to asset tracking.

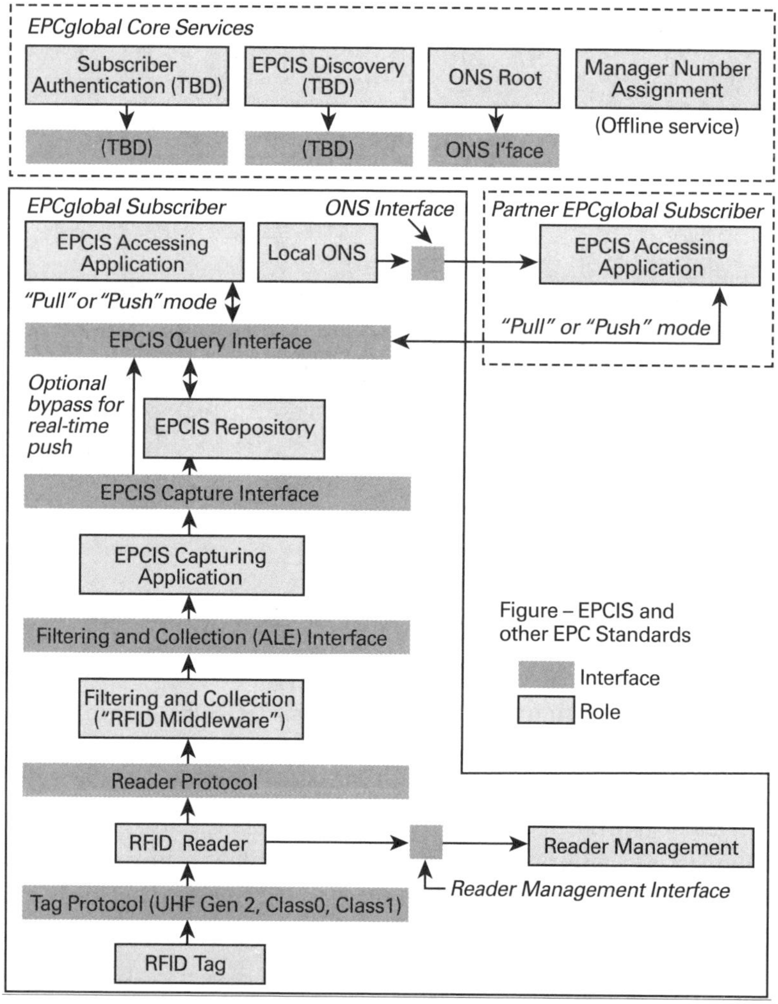

Fig. 3–5. EPCglobal Network layers

EPC Information Service (EPCIS)

The EPC Information Service makes EPCglobal Network–related data available in correct format to requesting services. Data available through the EPC Information Service may include tag read data collected from readers (for example, to assist with object tracking and tracing at serial number granularity) and instance-level data (such as date of manufacture, expiry date, and so on). Data could also include object class–level data such as product catalog information. In responding to requests, the EPC Information Service draws upon a variety of data sources that exist within an enterprise and are communicating back to the requesting party. When the EPC data is distributed across the enterprise, an industry may create an EPC access registry that will act as a repository for EPC Information Service interface descriptions. The EPCglobal Network EPC Information Service specification defines the protocol for accessing the EPCIS.

Object Name Service (ONS)

The Object Name Service provides a global lookup service to translate an EPC into one or more Internet uniform resource locators (URLs), where further information on the object may be found. These URLs often identify an EPC Information Service, though ONS may also be used to associate EPCs with Web sites and other Internet resources relevant to an object. ONS provides both static and dynamic services. Static ONS typically provides URLs for information maintained by an object's manufacturer. Dynamic ONS services record a sequence of custodians as an object moves through a supply chain. ONS is built using the same technology as DNS, the Domain Name Service of the Internet. The EPCglobal Network Object Name Service Specification defines the working of ONS and its interface to applications.

ONS local cache

The local ONS cache is used to reduce the need to query the global Object Name Service for each object that is seen, since frequently-asked/recently-asked values can be stored in the local cache. The local cache acts as the first port of call for ONS type queries. The local cache may also manage lookup of private internal EPCs for asset tracking. Coupled with the local cache will be registration functions for registering EPCs with the global ONS system and with a dynamic ONS system for private tracking and collaboration within the supply chain seen by each unique object.

Solution components

A fully instrumented RFID site is shown in figure 3–6.

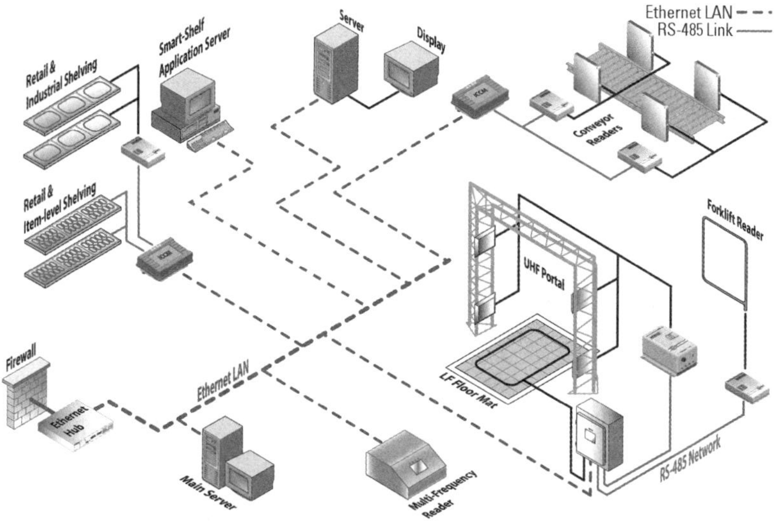

Fig. 3–6. Fully instrumented RFID site for total operational visibility

The bill of material for instrumenting this site is shown in the following chart:

Hardware	Est. Quantity
RFID Reader	1
RFID Antenna(s)	4
Mounting Kits, Cables	3
Site Survey	1
Passive Handheld RFID Reader and Mobile Station	2
Handheld Reader Accessories (Holster, Batteries, Adapters)	1
Tag Printers	2
RFID Tags	xxx,xxx
Server Platform	1
Other Cables, H/W	1
RFID Installation Support and Site Prep	80
H/W Contingency Reserve	1
Total	
Software	
Reader and Device Management Software	1
EPCIS Software	1
Total	

Excluding the cost for RFID tags, which is dependent on the number of items being tracked, enterprises should budget for $200,000 for preparing their site for RFID operations.

Using RFID for Improved Warehouse Management

Early usage of RFID has focused on bringing efficiencies within the warehouse management space. This section explains how warehousing can be significantly improved using RFID.

Shelf-life management, hazmat control with RFID

Read/write RFID tags can be used to store not only the item information that is typically contained in the bar codes, but also other industry parameters such as the expiration date, remaining shelf life, and hazmat classification. This information, critical to carrying out the logistics operation, can be housed right on the product, instead of just on remote databases. This can greatly streamline operations such as FEFO picking. For example, consider the following scenario:

1. Based on the remaining shelf lives of the products in the bins, EPCIS generates the most optimal FEFO pick sequence for satisfying the demand.
2. The picker takes the pick list to the right bins, pulls out the products, and fulfills the order.
3. If the picker has picked a product out of the FEFO sequence, this can be instantaneously detected by EPCIS communicating with RF readers either carried in the form of a mobile handheld device by the picker or as more permanent fixtures at the QA stage as shown in fig. 3–7.
4. A pick exception report can be generated that can redirect the picker to go back to the bin and rectify the situation.

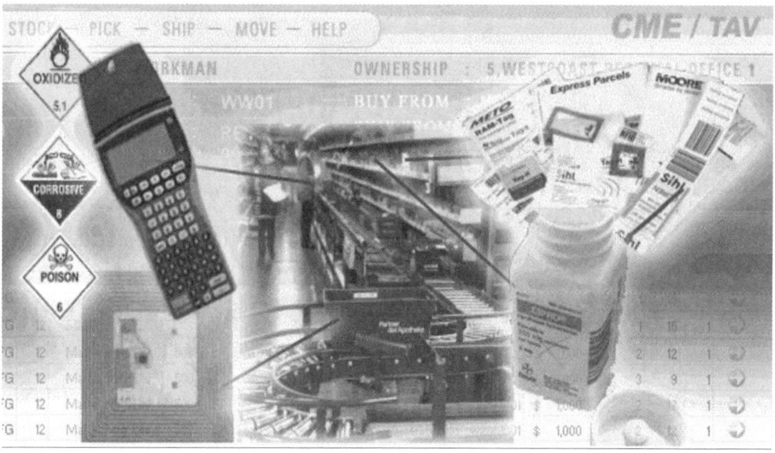

Fig. 3–7. Shelf-life management using RFID infrastructure

Similarly, better hazmat control can be achieved by affixing RF tags in addition to the hazmat labels for identifying and classifying the items as mandated both by OSHA (Occupational Safety and Health Administration) and DOT (Department of Transportation).

Powerful lot control

Lot tracking can be dramatically improved using RFID infrastructure. By using read/write tags on the products that contain not just the item information, but also the lot information (fig. 3–8), the movement history of the product can be recorded on the tag itself. This method of lot control is preferable to having to reconcile and join numerous disparate database queries across different segments of the supply chain. For example, the RFID tag can record the following item and lot information:

- Item number
- Lot number
- Vendor number
- Vendor lot number
- Manufacture date
- Carrier number
- Truck ID
- Receipt date
- Expiration date
- Retest date
- Lot status

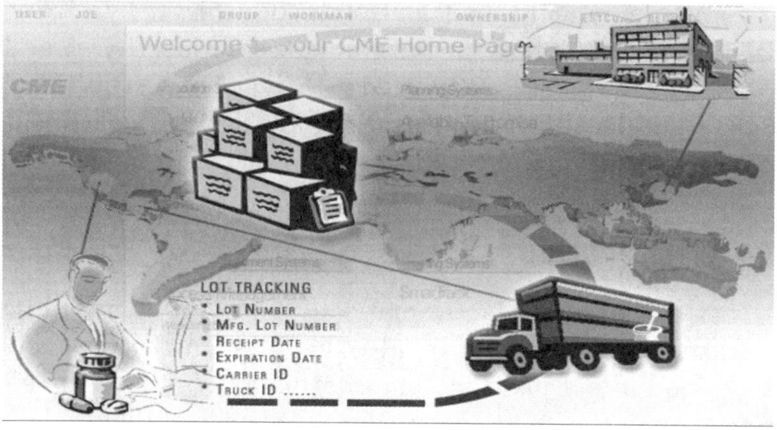

Fig. 3–8. Effective lot control by associating critical lot tracking information with RFID tag

In the past, this information only resided in the backend databases. As a result, all lot control was based on manual comparison of the items to a printed list. With this information also present on the tags, an RF reader can immediately read the entire lot history right off the products. Identification during a product recall becomes simply a matter of scanning the products in the bins. EPCIS software drives the readers and identifies the items in the bin that have been recalled.

Better warehousing and logistical operations

RFID infrastructure can eliminate or dramatically reduce problems that traditionally plague storage facilities such as misstocks, mispicks, and misships. With RF tags attached to the inventory items and EPCIS software managing the RF events (fig. 3–9), these errors can be easily caught, corrected, or fully eliminated. For example, consider the following misstocking scenario:

1. The worker accidentally stocks the product in the wrong bin.
2. The RF reader present in the bin reads the RF tag of the product.

3. EPCIS knows the product rightfully belongs in another bin, and thus it generates an exception event.
4. The worker is notified in real time via a buzzer or a notification to his PDA-based work list or at a fixed time period via an exception report.

Next a mispick scenario can be examined:

1. The worker picks out the wrong product from the bin. The product can either be the wrong type or can even just be out of the FEFO pick sequence.
2. A handheld RF reader carried by the worker reads the RF tag on the product.
3. The RF event is captured by the EPCIS software that immediately performs matching and discovers this error.
4. The exception is caught in real time by the EPCIS software, and the worker is notified on his or her pager or PDA about the problem and steps to remedy.
5. The problem is stopped before it snowballs into a larger issue.

Fig. 3–9. Streamlining warehousing operations with RFID infrastructure and EPCIS

With EPCIS software also running at the shipper or supplier, no mispicks occur there either, resulting in no misshipments at the receiving end. Thus operational errors are dramatically reduced on the floor and during transportation, resulting in physical and virtual inventory worlds that are closely in sync.

Countering counterfeits and thefts

RFID infrastructure can make a huge difference in the areas of counterfeit detection and theft prevention. Using the native RF capabilities of EPCIS software, the information on RF tags can be completely encrypted, adding a significant deterrence to counterfeiting. Moreover high-security RF tags can be employed that are perfect for anticounterfeiting. These tags store encoded data, using a rolling count both at the receiver and the tag and a two-stage read process. This almost makes it impossible to counterfeit the tag. These tags are being widely used in the ignition keys of new cars.

Similarly with RF readers installed around the distribution and transportation facilities, movement of the product within and across the facilities can be efficiently tracked using EPCIS software. Appropriate alerts (fig. 3–10) can be delivered when a product goes through a fixed checkpoint (such as an access door). Security personnel can be quickly notified on their pagers or PDAs and then dispatched to the right locations.

Fig. 3–10. Triggering alerts via PDAs for items removed illegally from the warehouse

Using RFID for In-transit Visibility

In-transit visibility is achieved through a combination of RFID technology and middleware infrastructure. This section will examine how key RFID technologies in combination with EPCIS platforms can provide item-level visibility and management of goods flowing through the utility supply chain. The major requirements of a solution for in-transit visibility, tracking and response, fuse innovative processes with available technologies and integration.

Innovative processes must include the following:

- **Leading supply chain practices.** These practices must be assembled and tailored to the specific situation, for example, one that considers both the vendors's supply chain objectives and the retailer's needs.
- **Change management.** Change management is necessary to ensure that the organizations "buy in" to the processes and enabling technologies and implement them fully.

The technology requirements must include the following:

- **Multisite visibility at the corporate level.** Companies and organizations with distributed operations often have good local visibility but minimal to no visibility at the corporate level.
- **Ability to support corporate-wide ad hoc queries.** Companies and organizations often do not have the right data structures to effectively query, much less analyze, distributed operations.
- **Business-level communication across sites rather than data-level communication.** Most distributed processes, such as resolving a misshipment, involve multisite, multiperson coordination.

- **Exception-based management.** Not only must the right exceptions be generated but also they must be routed to the right personnel.
- **Robust messaging support.** To ensure that goods can be tracked across locations or across states (such as in manufacturing or by location in a retail store), messaging support needs to include the following functionality:
    - Synchronous, asynchronous
    - Request-response
- **Prevent flooding the network with tag events.** To handle a fully distributed solution that is generating potentially millions of events per second with current approaches would overload a network. The right solution must process the majority of events locally yet be able to forward the right events to other systems in real time.
- **Leverage existing hardware/networking infrastructure.** To provide payback in a minimal amount of time, a solution must leverage the existing investment in hardware, software, and networking.

These requirements also must be combined with the following:

- **Robust wireless network infrastructure.** This infrastructure must provide visibility not only in transit but also within the organization's facilities.
- **Location devices and tags.** Location devices and tags must provide the location as well as other pertinent information about the goods. Chief among technologies to provide this functionality is RFID.
- **Distributed event-driven software infrastructure.** The software infrastructure must be RFID enabled, support one to many wireless

networks, integrate among enterprise legacy systems and ERP/ERM systems, and support applications that monitor goods.

Rather than address all aspects of the solution, this discussion focus, for the sake of brevity, will be on some of the key aspects of the enabling technologies.

Using passive RFID for in-transit visibility

Passive RFID brings strong capabilities of providing in-transit visibility at the item level. Benefits of using passive RFID include the following:

- No line-of-sight communication is required for reading.
- Tags have read/write capability, with ample storage for a variety of information.
- RFID works even under extreme environmental conditions.
- Tags are becoming increasingly affordable and smaller.
- A number of configurations support almost any good and security need.
- Anticounterfeiting can be implemented easily and cost effectively.

Read/write RFID tags can be used to store not only the item information that is typically contained in the bar codes, but also other industry parameters such as the expiration date, remaining shelf life, and other information. This data, which is critical to carrying out logistics operations, is housed right on the product, instead of just on remote databases. This can greatly streamline operations such as FEFO picking. For an example, the following lot control scenario can be considered.

By using read/write tags on the products that contain not just the item information, but also the lot information, the movement history of the product can be recorded on the tag itself. Again, this is preferable to having to reconcile and join numerous disparate database queries across different segments of the supply chain. For example, the RFID tag can record the following item and lot information:

- Item number
- Lot number
- Vendor number
- Vendor lot number
- Manufacture date
- Carrier number
- Truck ID
- Receipt date
- Expiration date
- Retest date
- Lot status

In the past, this information only resided in the backoffice databases. As a result, all lot control was based on manual comparison of the items to a printed list. With this information also present on the tags, an RF reader can immediately read the entire lot history right from the products. Identification during a product recall becomes simply a matter of scanning the products in the bins. The software must drive the readers and identifies the items in the bin that have been recalled or have expired.

Global and local wireless communications

The infrastructure for tracking assets, both globally and locally, is depicted in figure 3–11.

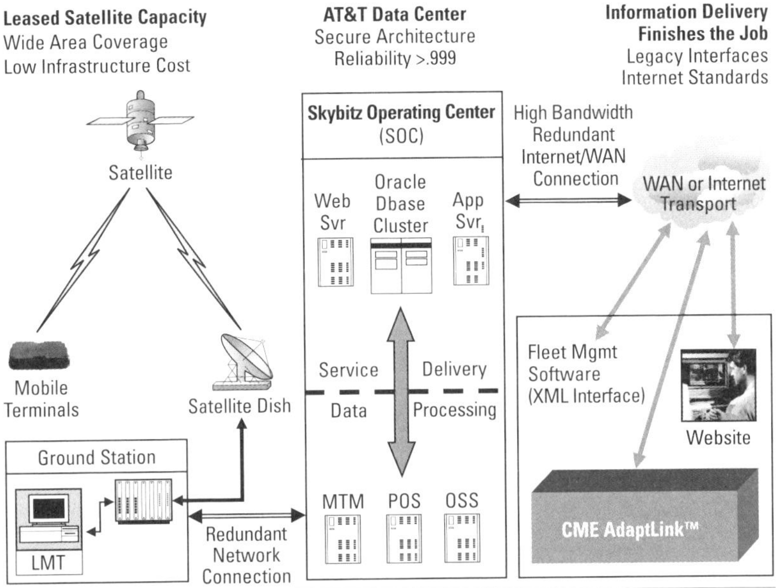

Fig. 3–11. Infrastructure for tracking assets locally and globally

Assets are either tracked via reading at key checkpoints or via transmitting to satellite at key intervals. The generated events are then transferred to the data center and transmitted to the EPCIS software, which then provides a unified view of the goods as well as their current location and status.

A satellite tracking system provides the ability to track the exact location of vehicles and convoys. The latitude and longitude locations of trucks, trains, and other transportation assets equipped with a transceiver are transmitted periodically via a satellite to a ground station. Some systems also provide two-way communications between a vehicle operator and a ground station for safety, security, and rerouting. Satellite tracking uses a cellular or satellite-based transmitter or transceiver unit to communicate positional information, encoded and text messages, and emergency messages from in-transit conveyances to the ground station. Transceiver-based technologies also permit communications from a ground station to the in-transit conveyance. A user can compose, transmit, and receive messages with small handheld devices or with units integrated with computers.

Yard- or facility-based wireless communications

The infrastructure for tracking goods within a yard or a facility is depicted in figure 3–12.

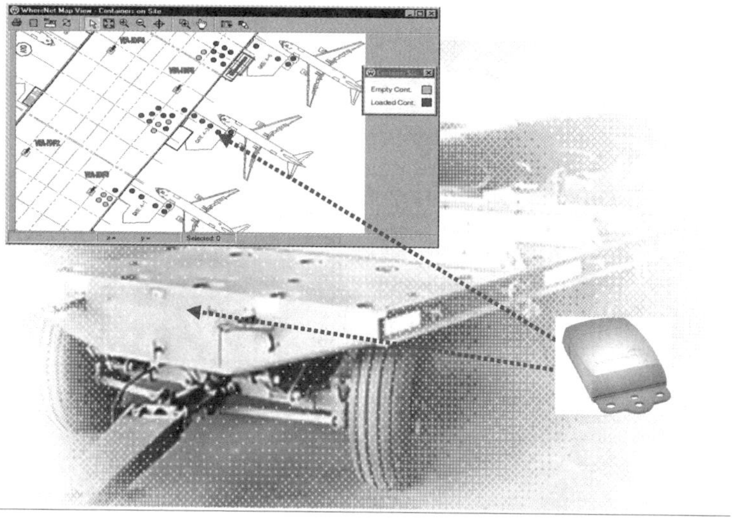

Fig. 3–12. Real-time location using active RFID on the tarmac

Typically, a satellite signal is blocked within a yard or facility by metal structures. So, a local wireless infrastructure provides visibility in these conditions. EPCIS software can be used to track individual trucks, their loading status, and the goods already loaded. EPCIS software can interface with a yard management system for scheduling loading and unloading as well.

Work orders, including maintenance, can also be transmitted wirelessly using the solution's wireless support for handheld devices. This maximizes the productivity of the direct workforce.

Enabling software for in-transit visibility

A generic EPCIS software needs to be enhanced significantly to provide in-transit visibility and leverage the investments made in the entire spectrum of RFID tags. Some of the key functionality includes capturing events with RFID support, processing them locally and remotely using collaborative processes, and providing a global view to headquarters.

In general, the enhanced software captures more precise and timely data, adapts it to unique circumstances, and then handles it by either routing it to the right systems or by setting up a collaborative workflow. The major components of this software should include the following:

- **AIDC/AIT subsystem.** Provides unique bidirectional support for data capture technologies (e.g., RFID and bar codes)
- **Unified data model.** A data model that supports unique item information (UII) and allows drop-in integration of supply chain and legacy data.
- **Business/process management utilities.** Allows business processes to be flexibly modeled rather than hard coded.
- **Event subsystem.** The event subsystem takes batch input, such as a scan of an entire pallet of goods, and breaks it down into discrete events that can be processed separately.

Conclusion

It is evident that using the entire spectrum of RFID tags and enabling software can provide in-transit visibility to assets that are flowing across the globe. It is important to mention that the focus on the business process is critical for implementing in-transit visibility, track and respond capabilities. Without attention to the process, the initiative is not likely to succeed. It must encompass the collaboration with the remote site, joint visibility and alert management, procedures and escalations in managing events and responding to customer demands, and the ability to configure warehouses to minimize stockouts.

Summary

The benefits for using RFID for mobile asset tracking in the utility industry are numerous. The business benefits—making this an extremely sound and justifiable investment—include the following:

- Better shelf-life management and more accurate picking
- Improved and rapid response to customer demand
- More effective and straightforward lot control and traceability
- Effective counterfeit prevention
- Reduced theft and loss
- Extremely accurate system inventory information—down to six-sigma operational levels
- Reduced operational cost due to dramatic reduction in unfilled fillable demand and problem-receive
- Efficient buying and shipping cycles
- More efficient fulfillment plans
- Reduction in safety stocks without affecting service level

In summary, this will enable a lower cost, customer-responsive enterprise for the utility industry.

4 APPLICATION OF RFID AND WIRELESS TECHNOLOGIES FOR FIXED ASSET LABELING AND LOCATING

Introduction

It is important for water, wastewater, gas, and other utility companies to identify, locate, and maintain their fixed assets efficiently. This can help them improve asset management and security. Fixed assets include underground and concealed utility lines and pipes, gas and electrical lines, gas and electrical distribution hardware, canals and aqueducts, reservoirs, storage tanks, pumps, valves, and more. In many instances, fixed assets such as pipelines and electrical lines are concealed between floors of buildings and walls of erected structures.

This chapter will examine how RFID technology can be used by utilities to track, locate, and manage their fixed assets. This is important because lack of proper knowledge and the inability to locate these assets rapidly and with precision can result in damage to these assets during excavation for construction or repair. Such damage results in injury, possible loss of human life, expensive downtime, project delay, and liability. It is also important to have accurate and timely information on the location and status of these assets during an emergency caused by an act of terrorism or during natural disasters such as hurricanes, tornados, and earthquakes. This could help minimize damage and loss from gas leaks, prevent contamination of drinking water, and prevent people from getting electrocuted.

Existing Technologies

Existing techniques used by utilities for detecting such underground assets include the use of ground penetrating radar (GPR), metal detectors, and GPS-based systems.

GPR-based systems generate large amounts of data, much of which is unwanted and is not specific to the location of the line to be identified. This results in inaccuracies. The main problem with using GPR in locating underground assets is the inability to distinguish between the signals returned by an underground object of interest and those signals returned by other subsurface objects.

Another technology that has been used is the application of metal detectors. While metal detectors can detect metals, they cannot differentiate a metallic utility line from another piece of metal.

Faced with these challenges, utilities have started using GPS for locating these fixed assets. Existing maps are first digitized to generate coordinates for assets on the map and the data is stored in a database. This data is then accessed using a GPS-enabled handheld device that directs the user to each asset. Unfortunately, there are a number of limitations with this approach. The accuracy of most GPS devices is 3 to 5 m unless differential GPS is used. Unfortunately, differential GPS requires significant processing time at each location along the asset and therefore becomes expensive and impractical as a locating technology. Given this margin of error, locations identified with GPS also lack in repeatability. Additionally, GPS requires line of sight with satellites and does not function properly when there is shading from trees, buildings, or other objects.

The limitations of existing technologies demonstrate the need for an accurate and inexpensive solution to extend existing GPS-based solutions by addressing problems with accuracy, repeatability, and the need for constant line of sight with satellites. One can apply RFID and wireless technologies to develop a solution that provides the ability to detect, map, track, and locate the assets in real time and to extend the capabilities of GPS-based solutions. Additionally, RFID tags can be used to store historical data on maintenance and replacement, thereby improving the management of the assets.

Development of an RFID-based Fixed Asset Labeling and Locating System

Problem

One recurring problem is the damage to adjacent underground lines that occurs during excavation to access and repair a pipeline. This damage is due to the proximity of lines to other structures and utility lines, their colocation within the same vertical plane, and their deviations from the locations shown in the drawings.

A second problem facing field crews is the identification of similar assets. For example, when a crew is in the field, it may take a considerable effort to distinguish one valve along a pipeline from another in the vicinity. If there are parallel underground pipelines, the crew is sometimes unable to distinguish the lines once they have been excavated.

Even after the assets have been located (e.g., valves on parallel pipes), the crew has difficulty ascertaining when the last repair was conducted unless there is an ID number associated with the valve and the information on the repair has been entered into a database. In some instances, the information obtained from the database may appear to differ from the condition observed in the field (the valve appears newer or older than the date of repair stored in the database). This could be because the person entering information into the database is at an office and is not the same person who took notes in the field. Alternately, the repair and maintenance tasks may have been outsourced to contractors who collected the data, providing reports to the utilities that were then not properly updated in the asset management database. The field notes may be partially legible and errors arise when transcribing the information from the notes to the database.

Several of these problems can be precluded if the valve ID and the history of the repair can also be programmed into the memory of the RFID tag. Having an automated system that allowed repair and maintenance data to be updated on the servers simultaneously as the tasks are performed would bring about efficiency and also prevent these errors.

Solution

As previously discussed, there are three main problems—identification of assets, locating them rapidly and with precision, and having access to accurate history of maintenance and repairs. RFID technology can be applied to solve these problems effectively.

Asset labeling for unique identification

In order to be able to identify assets uniquely, they can be labeled with an RFID tag. An RFID tag has a unique serial number and optionally some memory. The RFID technology to use for such labeling would depend on the environment. If a label is to be applied to an asset that is easily visible and reachable, 13.56 MHz–based passive tags can be used for labeling. The 13.56 MHz passive RFID technology has no issues with water and metals, unlike 951 MHz, but has a small range (6 in. for a handheld RFID reader). On the other hand, 951 MHz passive RFID tags have a higher range but do not work well in the presence of metals. Active RFID tags can provide a higher range and are also not affected by water and metals, but they are more expensive.

After an asset is labeled, its unique identifier is stored in a database on the server along with additional information. This additional data could include the asset type, make, date of purchase, its location (absolute coordinates or location relative to another asset, landmark, or a primary label on a landmark as discussed below), and installation and maintenance history. If the RFID tag has its own memory, some of this information may be stored in the tag also to provide quick access to the data and a level of redundancy.

Asset location

If an asset is labeled and its approximate location is known, it should be possible to locate it and then read the RFID-based identifier for validation. This process of visually locating an asset using a reference landmark (such as a corner or door in a plant facility) works if there are not too many assets of the same type in the area and the assets are easily visible and accessible.

Sometimes, locating an asset inside a large facility or determining the exact location of an underground asset that is not visible can be difficult. It may therefore be necessary to have other identifiers that can lead to the asset or multiple points on an asset. Following that, the label or labels on the asset can be used to validate and uniquely identify it. To distinguish between these two types of labels, the labels that are placed directly on or above an asset as will be referred to as *secondary labels* or *secondary markers* (or *direct labels*). The labels that are used to locate these will be referred to as *primary labels* or *primary markers* (or *indirect labels*). Such labels (markers) embedded with RFID chips installed at the corners of buildings, on fixtures, or at locations along the edges of tanks, etc., are shown in figure 4–1.

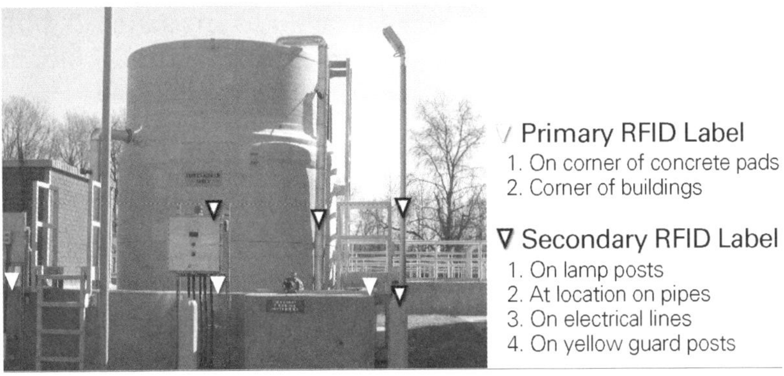

Fig. 4–1. Illustration of the use of primary and secondary RFID labels to label and identify assets at locations where they are visible

The primary labels are used at readily accessible points as shown in figure 4–1 (such as the corner of a tank or near the light switch inside a room). It stores the coordinates and description of secondary labels. It may also store coordinates of points along a pipeline. The method of storing coordinates is discussed later. The primary labels are used to locate the secondary labels.

The secondary labels are located on the asset or in the vicinity of the asset that needs to be accessed. The label ID and information stored in the memory is accessed through a handheld device.

The primary and secondary labels may also be used to access the coordinates of various points along a hidden pipeline relative to the labels (fig. 4–2). The coordinates of different points of the hidden pipeline are stored in the server database, either as absolute coordinates, relative to a landmark and its primary label, or as coordinates of secondary labels. Information on the points where adjacent lines cross over each other is also stored. The spacing between adjacent points along a pipeline stored in the database is based on several parameters, such as the following:

- The location of adjacent ends of each straight section of pipeline.
- The locations where adjacent lines cross over each other or come close to each other.
- The number of points required to describe the curvature of the pipeline at its bends. (The number of points selected should fully describe the arc.)

During a field visit, by reading one or more primary labels in the area using an RFID reader, it is possible to obtain the location information of different points on the hidden pipeline and the secondary labels. This information can then be used to locate the points of the hidden line using a simple laser-equipped survey instrument. The methodology is described next.

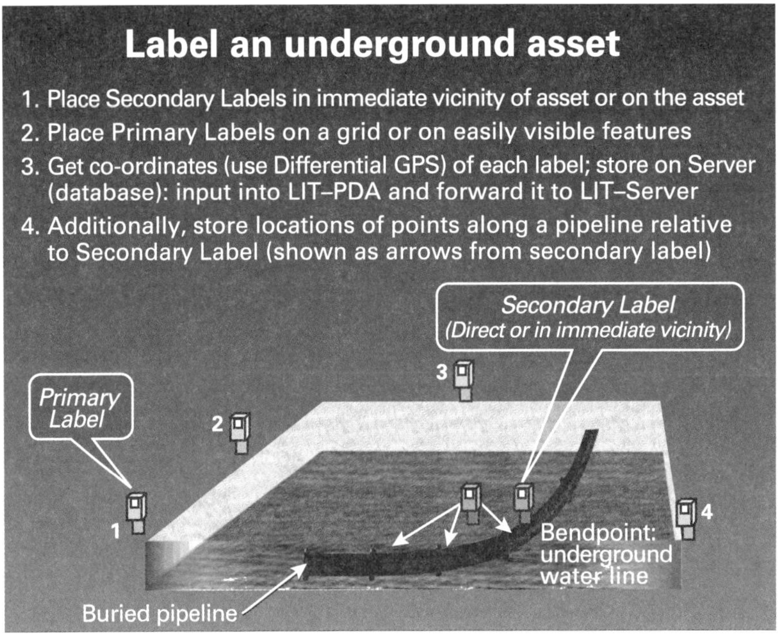

Fig. 4–2. Labeling an underground asset with primary and secondary RFID labels to help with its location

The RFID chips used as part of the labeling system are selected based on their ability to survive over an extended period of time (20 to 50 years). The locations of the markers should be such that they are permanent (not damaged or moved during maintenance or construction). Only when major construction is done would the markers and labels be removed and replaced by additional locations that are labeled. The labeling system must have a level of redundancy such that when a point is lost (due to storm damage, construction damage, a chip goes out of service, etc.), the location of the pipeline can be determined by the coordinates derived from other markers in the vicinity.

Repair and maintenance

Once an area or facility is properly labeled with primary and secondary labels, different techniques can be used to locate assets (under or above the ground) and to provide instant access to relevant information in real time during a repair or maintenance operation.

To determine information on the status of a fixed asset, information on each inspection and repair can be stored (and updated) by the field crew in the memory of the RFID tag (microchip) placed on the secondary label. If wireless access to the server is possible, the information can also be stored on the server. For example, the memory on the chip can be used to store the last repair or service date on valves in a pipe vault. Where wireless access is not possible, the data is entered in a handheld device in the field and synchronized with the enterprise system at the end of the day. The operator can log into the database using his or her ID or an RFID-based access card. When the pipe vault is accessed in the future, the crew can obtain a confirmation of the maintenance record (such as the last maintenance date) by using a handheld device to read the information stored in the memory of the RFID chip.

Figure 4–3 shows the architecture of a software system that can be used to implement solutions for identifying, locating, and maintaining history of repair data. Information on the assets, primary and secondary labels, and maintenance history are stored in the server database. This information is updated by operations staff using a handheld device, either in real time if a wireless link to the server is available or by collecting the data locally during the operation and then synchronizing it with the server database subsequently. The handheld device also includes an RFID reader to read primary and secondary labels during operations.

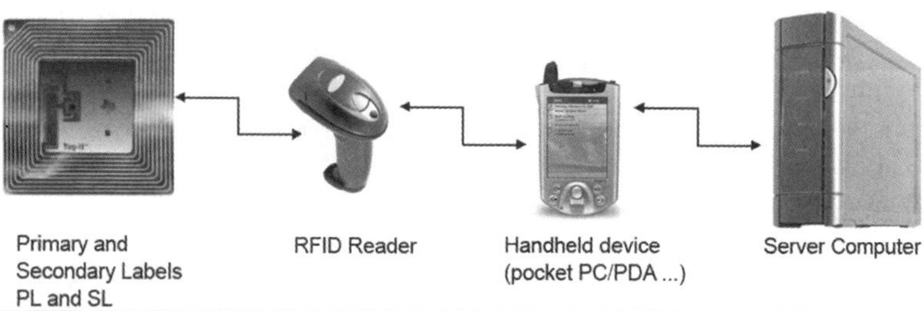

Fig. 4–3. Layout of components of the system

Example of an implementation

This discussion will now consider an example that uses RFID technology to track and locate the survey markers (pins) and the underground utility lines adjacent to the creeks (fig. 4–4). Instead of creeks, one may consider a similar application along a utility's right of way, or along the route of a natural gas pipeline. The intent is to be able to locate the asset faster, with precision, and to overcome the limitations faced by the utility with GPS and metal detectors.

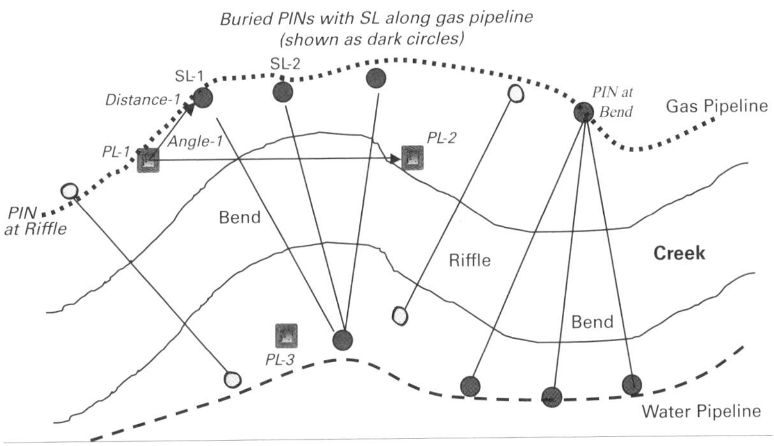

Fig. 4–4. Locating pipelines and fixtures along a terrain with primary and secondary RFID labels

Figure 4–4 shows the application in a creek. Primary labels (which are embedded with active or passive RFID tags) are placed in the vicinity of the permanent fixtures (tree trunk, concrete vault, or a structure that is readily visible). They are located above the ground and in areas where they will not get wet or damaged. The secondary labels are embedded on the survey pins and on fixtures (such as a valve on a pipe). The coordinates of features along the creek and assets running alongside the creek (such as location of a storm drain outfall at the creek or a buried utility line alongside the creek) are assigned relative to the primary labels. If the secondary labels are offset from the fixtures, they are also assigned relative to the secondary labels.

Figure 4–5 shows how the overall system works. The handheld RFID reader reads the content of a primary pin. The handheld device communicates with the server computer to send and receive data over a wireless network when a wireless network is available at the location. The software has a server component called the *s-sw* and another that runs on the handheld device (called *h-sw*). The h-sw communicates with the RFID reader and the server, using wireless Bluetooth and wireless GPRS technologies, respectively.

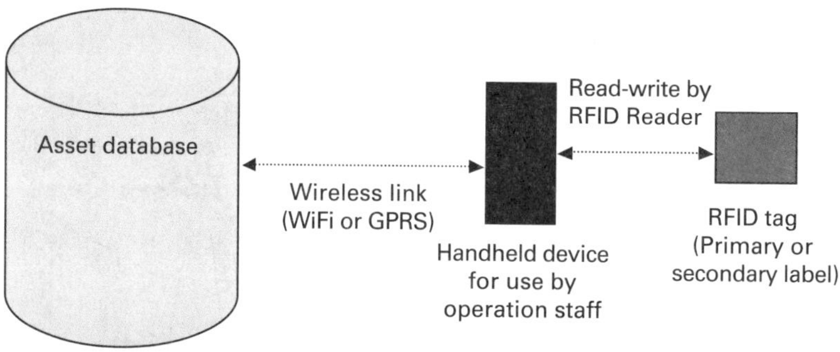

Fig. 4–5. Schematic showing communication and transmission between the handheld device (read/write RFID reader), the RFID tag, and the asset database

If a wireless link is not available at a location, appropriate data snapshots may be downloaded to the handheld device (using the software running on the handheld) before going to the field. The handheld device should be selected based on the amount of data storage required. Sometimes a utility has a significant amount of its assets in remote areas (with no wireless access) where there is significant shading from trees (no GPS signal). In such cases, the handheld device should be able to download and store the coordinates of secondary labels and the coordinates of points along a pipeline in the vicinity of the work area to be visited on the particular day.

Labeling the assets. The first step involves identifying the best locations for the primary labels in the plant or field and obtaining the differential GPS coordinates of the primary labels. The next step is to measure the locations of each of the secondary labels (in this example, each survey pin) relative to a pair of primary labels. This involves using two nearby primary labels, one as the *base primary label* and the other as the *reference primary label.* A survey instrument (such as total station or laser) is placed on the base and focused on a prism on the reference to obtain a *reference traverse.* As shown in figure 4–4, the location of a secondary label is obtained by placing a prism on the base primary label PL1 and obtaining the following:

- The angle turned from the reference traverse
- The distance from the base primary label

Figure 4–4 shows the angle 1 and distance 1 for secondary label SL1. Angle 1 represents the angle by which the laser needs to be rotated from the reference traverse [i.e., the angle between the line joining the base primary label PL1 and reference primary label PL2, and the line joining the base primary label PL1 and secondary label SL1]. The distance 1 represents the distance of secondary label SL1 from the base primary label PL1.

If the primary label is a corner of a pad (fig. 4–1) on which a survey instrument cannot be placed, one may embed a metal tag or pin on the ground at a point in its vicinity on which the instrument can be placed. This point is recorded as an offset to the primary label. The coordinates are stored relative to this offset.

This process may be repeated to store information of points along a pipeline that are not marked with a secondary label. This is shown in figure 4–2.

The locations of the primary and secondary labels are stored on the server (see fig. 4–4) in its location database. The (angle, distance) data for a pin is converted to absolute GPS coordinates before it is stored in the location database.

For existing underground assets, the location information of an existing underground utility line is obtained from existing maps. The maps are digitized and orthorectified to import them into a GIS system (such as ArcView[1]). The possible locations of the primary and secondary labels are identified and placed on the maps. The relative angles and distances are obtained from the map. The coordinates of locations along the underground line are then stored on the server.

During the installation of the primary and secondary labels in the field, the locations of the utility line may be field verified if the line can be accessed at certain points along its length. The relative distances and angles between the primary and secondary label are field verified and the database is updated. If the line is exposed during a future construction project, the relative distances and angles are verified and updated. This process of obtaining and updating the coordinates of the primary and secondary labels and points along the underground utility line is called *asset labeling.*

Locating the assets. At a later date, someone may have to locate the secondary labels and points along the pipeline in figure 4–3. This may be at the beginning of a new construction project. The first step is to use a GPS (the handheld may have GPS capabilities), a map, or a known landmark to arrive at the approximate location of the primary label. The label is identified visually. The system may be enhanced by including a digital display on the handheld device that helps with the location of primary labels. For the digital display, the current GPS location or the name of the landmark is input into the handheld device software to obtain the list of primary labels in the area (displayed as nearest one first, farthest one last).

Next, the RFID reader is used to locate two of the nearest primary labels. As explained previously, one of the two primary labels is called the base primary label (base PL) while the other is called the reference primary label (reference PL). The GPS coordinates of the base and reference are displayed by the handheld device software. A survey instrument (possibly a laser instrument such as a Contour XLRic from LaserCraft, Inc.[2]) is placed on the base and focused on a prism on the reference PL to obtain a reference traverse. By selecting a particular primary label as the base PL, the software on the handheld (h-sw) displays the locations of secondary labels (and points along an underground line) that are within a certain specified range. For each secondary label (SL), the locations are displayed by the h-sw as angles and distances. The angle is the angle by which the laser needs to be rotated from the reference PL to point to the SL, and the distance is the distance of the SL from the base PL. The laser is turned accordingly to point at the SL. A person moves the prism along the line joining base PL and SL to measure off the correct distance. This enables the location of different points on the ground surface that are directly above an underground utility line.

If an SL is lost, this process can be used to locate where the SL should have been present and to place an SL (a replacement RFID tag) at the desired position.

Summary

The RFID-enabled system described in this chapter allows engineers, contractors, field inspectors, and maintenance staff to locate and detect underground and other fixed assets (including water, wastewater, gas, etc.) in a plant or city area with accuracy. It adds precision to GPS-based solutions whose accuracy and repeatability are not always adequate.

The solution works in all areas including those where a GPS does not work because of lack of direct line of sight with the satellites (because of shading between buildings or from trees). For the same level of accuracy, the solution is rapid because it avoids the signal time and the processing time required to obtain a differential GPS reading.

In evaluating RFID for fixed asset management, one can look at the solutions as a two-step process. The first step involves labeling at visually accessible points, both inside and outside buildings. Fixtures are labeled such that they can be identified rapidly and definitively and their maintenance history can be accessed. This is especially important in utilities where staff retirement leads to lack of prior knowledge about the types of fixtures inside a building, their locations, and the history of replacement and upgrades. The RFID-based labeling system enforces a discipline and enables better asset management.

The second step involves the upgrade of the program to help detect hidden fixed assets. This can be viewed as a long-term risk management investment. This RFID-enabled application reduces the time spent in locating underground and other fixed assets, preventing accidental breakage of utility lines and disruption during repair or construction. It also helps avoid the unplanned costs associated with the latter. In addition, the solution can be used to rapidly locate project boundaries and locations to collect critical survey measurements and to denote the position of utility lines along walls of buildings and industrial facilities. A location database comprising such information, along with the maintenance history of the assets by location, can be very valuable to utility companies.

References

1. ArcView, ESRI, Redlands, CA. http://www.esri.com.
2. LaserCraft, Inc. http://www.lasercraftinc.com.

5 RFID AND WIRELESS APPLICATIONS FOR ASSET MANAGEMENT IN THE WATER UTILITY

Introduction

The water utility industry was created in the 1930s to service the water supply and wastewater treatment needs of a community. The operations were part of a municipality. The service area of several utilities grew over time to include a region, a district, or a county. Together with the growth in the service area, the ownership and management has changed with time. The large utilities are operated as public or private entities.

The water utility is funded by water rates. The rates cover the cost of operations, capacity improvement, and infrastructure replacement. In the United States, the rates are increasing faster than inflation because of challenges with replacement of aging assets, difficulties in obtaining new raw water resources, and higher treatment and reuse requirements. Utilities are adopting higher levels of automation and staff and resource management to control the increase in rates. RFID and radio transmission technologies offer opportunities to improve asset management, monitoring, and tracking to help achieve these objectives.

General layout of assets

The assets of a water utility can include all or part of the water supply and wastewater treatment systems (fig. 5–1).

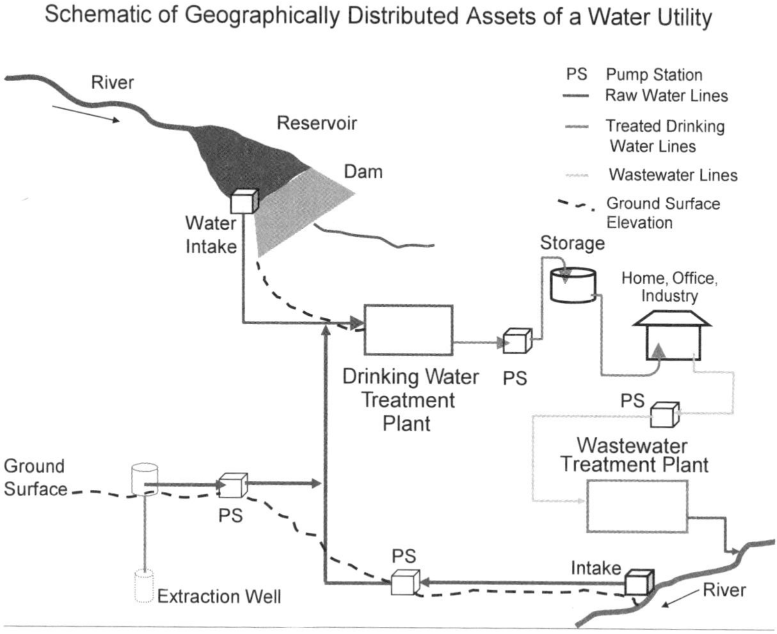

Fig. 5–1. General layout of a water utility

Within the water supply, the utility may have water pumping and treatment systems. These systems pump the water from various sources such as reservoirs, groundwater aquifers, and rivers to treatment plants. In developed economies, the treatment plants are designed to treat the water to a quality that is acceptable for drinking water. The water leaving the treatment plant enters storage and distribution systems. These are used to convey the treated water to homes and businesses.

After the water is used for washing, cleaning, or other services, it leaves the homes and businesses as wastewater. The wastewater is sent through a collection system that consists of sewer lines, sewer pipes, and pumps. The wastewater flows by gravity or is pumped to wastewater treatment plants. Here, the wastewater is treated to remove pollutants. The treated wastewater is discharged to a receiving water body such as a river or a lake, sent for irrigation, or recharged to a groundwater aquifer.

In addition to water and wastewater treatment, the water utility manages assets that are held in inventory in warehouses and mobile assets such as vehicle fleets that are used to service other fixtures. The water utility also manages specialty equipment such as tools and handheld computers used by service crews during the maintenance of treatment plants and the distribution system.

Structure of this chapter

This chapter is structured in two parts. The first part offers a description of the facilities within the water utility industry, including a description of the infrastructure for wastewater treatment and water reuse functions. The second part details the challenges with the management of the assets and the solutions that can be implemented. For fixed assets, these challenges are in locating, maintaining, tracking, and monitoring the usage and the condition of the assets, and their security. For mobile assets, the challenges are in tracking, shrinkage, utilization, and management of the mobile assets and products. The solutions are based on the RFID and wireless technologies presented in chapters 1 to 4. Measurable improvements may include a decrease in the rate of service disruptions, better utilization of resources, higher levels of monitoring of the infrastructure and the treatment process, better security for the infrastructure, and higher levels of customer satisfaction.

Description of Assets within a Water Utility

The infrastructure managed by water utilities can include all or parts of the water supply, distribution, and wastewater systems shown in figure 5–1.

Water supply

A water supply system includes reservoirs and dams to store the untreated (raw) water (fig. 5–1). It includes intake structures on rivers or groundwater wells that pump the untreated water to aqueducts and pipelines that convey the raw water to the water treatment plants. In some instances, the reservoirs for water supply may be shared with hydroelectric utilities. The utility has easements and right of ways for the pipelines and other infrastructure. These easements may be shared with other utilities such as electric, gas, and cable utilities.

Water treatment

A water treatment plant treats the raw water that is pumped to it from the water sources shown in figure 5–1. The typical water treatment plant removes the sediment present in the water by a process called coagulation (fig. 5–2). In this process, chemicals such as alum are added to coalesce the dispersed sediment (clay and silt) and settle them out in settling tanks (also called *sedimentation tanks* or *clarifiers*). The water is also treated to remove foul taste or odor, filtered to remove residual particulate matter (filtration), disinfected to kill pathogens (disinfection), and pumped to a distribution network that services the households and industries. The water treatment plant generates residual solids (sludge) from the sedimentation process. This has to be dewatered and hauled to land where it can be added to the soil or to landfills.

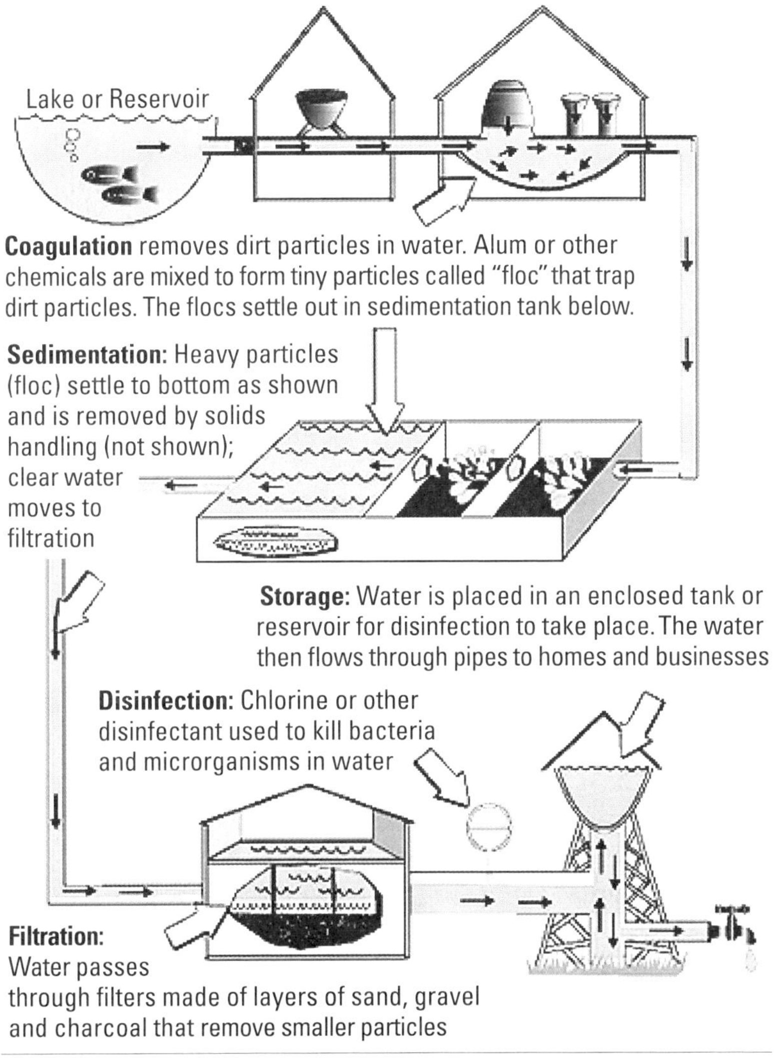

Fig. 5–2. Schematic of a water treatment plant. *Source: AWWWA Drinking Water Week Blue Thumb Kit*

The infrastructure within a water treatment plant includes the following unit processes. They are arranged in the sequence shown in figure 5–2.

1. Screening systems (if required) to remove coarse debris before raw water is pumped to the treatment processes downstream.
2. A pumping system to lift the water up to the treatment tanks.
3. Piping to convey water, chemicals, and solids between tanks and to various treatment processes.
4. Tanks with mechanical equipment such as mixers, flocculators, sedimentation basins, scrapers, and pumps that are part of the coagulation system.
5. Chemical feed systems to add coagulants (such as alum) to remove particulate material; calcium, magnesium, and sodium hydroxide to adjust pH and alkalinity; and oxidants such as permanganate to oxidize dissolved metals (such as excess manganese and iron) and precipitate them as sludge in the treatment tanks. (For example, ferrous iron is oxidized to a ferric form that is less soluble; the ferric hydroxide forms a precipitate that settles out in the sedimentation tanks, where it is then removed as part of the sludge pumped from the bottom of sedimentation tanks.)
6. Filtration devices (sand filters, membrane filters, etc.) to filter out particulate matter that remains in the treated water leaving the sedimentation tanks.
7. Disinfection systems that use ozone, a hypochlorite chlorine solution, or gaseous

chlorine to disinfect pathogens that may be present in the treated water:

- *Disinfection with chlorine gas.* This disinfection method requires storage tanks and protection systems for the use of chlorine gas. Disinfection with a sodium hypochlorite solution requires liquid feed pipes and pumps. Both disinfectants require mixing equipment to disperse (mix) the chemical into the treated water.
- *Disinfection with ozone.* The ozone generation system includes a high-voltage generator (to convert oxygen to ozone) and an ozone gas injection system to disperse the ozone into the treated water.

The use of certain chemicals such as gaseous chlorine requires stringent monitoring for worker safety. Chlorine leaks can be a hazard to workers at the plant and to residents in the vicinity. Ozonation systems have stringent safety requirements in the vicinity of the high voltage electrical generators.

Modern water treatment plants have high levels of instrumentation for plant automation and online monitoring. Several plants use distributed control systems such as a supervisory control and data acquisition (SCADA) system. Parts of the plant and processes are operated remotely. The RFID and wireless technologies, when applied, need to be compatible with the existing systems to enhance the monitoring, lower the cost of operation, improve the safety, or reduce the risk.

Water distribution

Once the treated water leaves the water treatment plant, it is fed by gravity or pumped to the water distribution system. The distribution system has a set of pipelines, pumping stations (also called *lift stations*), and valves that convey the water to intermediate reservoirs or storage tanks. From these tanks, a network of pipelines and valves conveys the water to a distribution system. At the end of the distribution system are water meters and faucets located at homes and businesses.

The pipelines that are part of the distribution system run under roads, alongside creeks, and through various right of ways that may be in public and private properties. Often, the pipelines are in the vicinity of other pipelines and conduits that convey natural gas to homes and electrical and fiber optic lines for a telecommunications network.

There are several aspects of the distribution system that lend themselves to the application of RFID and wireless technologies. The crews that service the network of pipes and fixtures need a reliable system to locate the underground assets that belong to the water utility and other utilities. They have to be scheduled based on the nature of the tasks and the availability of vehicles and tools to conduct the repairs. The staff who read the water meters face challenges with access to facilities. Improvements in their ability to read the meters remotely (from within 30 m or 100 ft) rather than by entering the properties would improve their productivity.

Additionally, there are instances where the utility would like to manage and balance the hourly consumption of water. RFID and wireless technologies can be applied to enhance the monitoring of flow rates on an hourly rather than monthly basis, record the data on a chip, and transmit the data over a wireless network. Flow monitoring along a pipeline, together with similar data recording and transmission capabilities, can help water utilities track losses from leaks in the pipeline and repair them in a timely fashion.

Wastewater collection

The water used in the household process or in the industrial (manufacturing) process is discharged through a second pipeline. This pipeline is completely isolated from the water distribution pipeline and is connected to the wastewater collection system.

The pipelines servicing a wastewater collection system form the wastewater (sewerage) network. Typically, these pipelines run at greater depths below the ground than do other utility lines (water supply, natural gas, or electrical). They drain by gravity to a local pumping station. The pumps at the pumping station pump the wastewater into a pressurized pipe that conveys it to a wastewater treatment plant (fig. 5–1).

A large wastewater collection system can have more than 100 intermediate pumping stations. It is necessary to have a monitoring system to monitor the status of the pumping stations (including equipment such as pumps, valves, and electrical units). Several utilities have installed a SCADA system with wireless transmission of information from the pumping station to a central location.

The pump stations have to be inspected and maintained by crews in response to alarms registered on the SCADA system and as part of the periodic maintenance. Additional inspection of the piping system is performed at the pump stations or at maintenance points (manholes) in the gravity sewers. These manholes are used to access the lines, insert cleaning devices, and clean out blockages or service the lines.

Many of the challenges with the maintenance and repair of wastewater collection systems are similar to those faced in the water, electrical, telecommunications, and gas distribution systems. There are assets that are hidden, distributed over a wide area, constructed over different periods of time, and located in close proximity to other assets. The maintenance crews have to cover a wide geographic area. RFID, wireless, and GPS technologies can help address these challenges if the location coordinates of the hidden assets and their maintenance history can be determined in the field with greater certainty.

Wastewater treatment plant and water reclamation/reuse

The wastewater treatment plant is designed to treat the wastewater and generate an effluent whose quality is suitable for discharge to water bodies such as lakes, streams, rivers, and estuaries, or for water reuse. Wastewater treatment plants can treat municipal or industrial wastewater. While a municipal plant treats wastewater that is principally from municipal sources (houses and offices), the industrial wastewater treatment plant services an industrial facility (e.g., pharmaceutical, beverage, food processing, chip making, pulp and paper).

As fresh water becomes a commodity whose supply falls short of demand, several utilities find it necessary to treat the wastewater to levels suitable for reuse (fig. 5–1). The level of treatment required at a wastewater treatment plant is commensurate with the type of water reuse. The types of water reuse practiced include: (1) agricultural reuse, (2) secondary water reuse, (3) indirect reuse, and (4) direct reuse. The levels of treatment required increase from reuse types 1 to 4.

Agricultural reuse is water reuse that requires primary and secondary treatment at a wastewater treatment facility with some degree of disinfecting and filtration. Primary treatment involves settling the coarse solids present in the wastewater by gravity in settling tanks or ponds (primary settling tanks). Secondary treatment removes a large fraction of the dissolved and colloidal organics that do not settle in the primary settling tanks. A schematic of the secondary treatment system is shown in figure 5–3. It includes screening systems to remove rags, grit removal to remove sand and coarse particles, primary settling tanks to settle coarse solids, and activated sludge tanks with aeration systems to grow bacteria to consume colloidal and dissolved organics (figs. 5–4 to 5–7). A secondary treatment system also includes secondary clarifiers to settle out the bacteria from the treated water and disinfection systems to disinfect the treated water prior to discharge. Additionally, it contains processes to handle the coarse solids and biosolids that are generated and processes to handle the chemical feed systems (fig. 5–8).

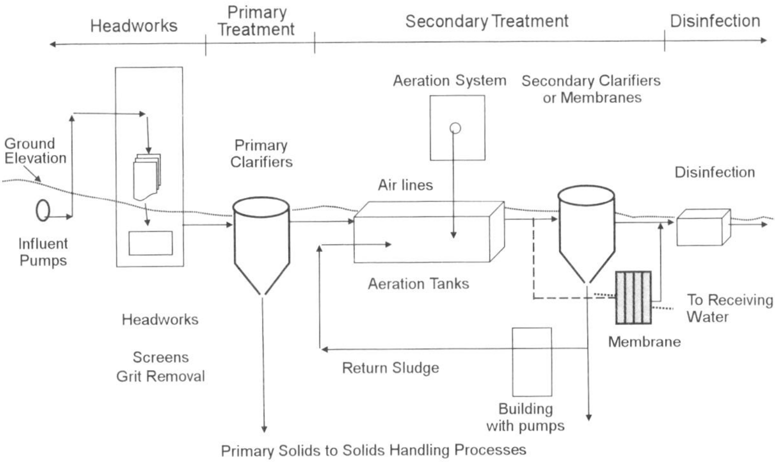

Fig. 5–3. Schematic of a wastewater treatment plant. Only liquid treatment process is shown. The biosolids handling process is not shown.

Fig. 5–4. An empty aeration (activated sludge) tank used for removing organics, nitrogen, and phosphorus from wastewater. *Source: Annapolis WRF, MD, 2006*

Fig. 5–5. Activated sludge tank in operation

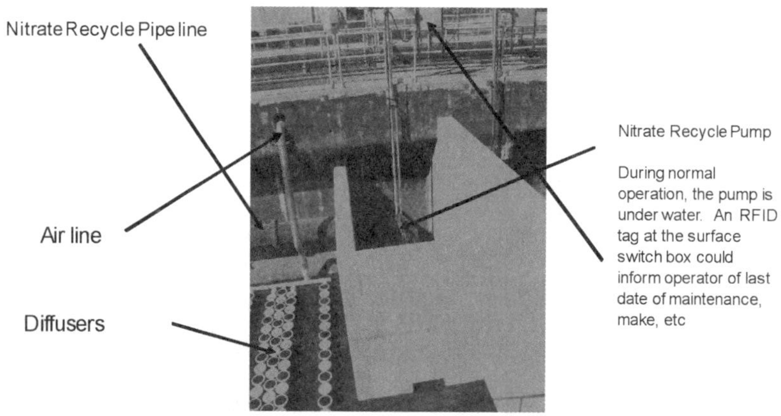

Fig. 5–6. Aeration pipes and diffusers inside an activated sludge tank

Fig. 5–7. Blowers to provide air to the aeration system of an activated sludge tank. The air inlet is from the roof intake (atmosphere) plus air vented through the Headworks building. The odorous air from the headworks is collected, pumped through underground air lines into the blower intake, and then pumped into the aeration tank where the bacteria oxidize the chemicals and remove the odors.

Fig. 5–8. Chemical feed system

Secondary water reuse is the reuse of water after secondary treatment. It is practiced by several water utilities. In this instance, the treated water is disinfected and sent to a parallel pipeline network that services lawns and toilets but is not used for drinking water. Its water quality is similar to raw drinking water before treatment at a water treatment plant. Secondary reuse reduces the demand on raw water sources by eliminating part of the demand from toilets and irrigation.

Indirect reuse is the third level of reuse for which the wastewater is treated to more stringent standards of tertiary treatment. This includes biological and chemical nitrogen and phosphorus removal, filtration to remove particulates, and in some instances, removal of trace organics in carbon adsorption columns. The water is then sent to a reservoir that serves as a drinking water source for a water treatment plant. An example is the Upper Occoquan Sewage Authority (UOSA/Millard Robbins) plant in Northern Virginia that treats the wastewater and sends it to the Occoquan reservoir. The reservoir is a drinking water reservoir.

Direct reuse is being practiced in a few instances. The disinfected wastewater is treated through membrane filters that achieve a level of residual solids removal similar to water leaving a water treatment plant. These may include membrane filtration using micro and ultra filtration followed by reverse osmosis. The water is disinfected further, if necessary, and is used as a water source for industries or as drinking water. Direct reuse has been implemented by the water utility in Singapore (NEWater) and by NASA (in the International Space Station).

Mobile assets

Tools and equipment. Each water utility maintains a variety of assets in one or more warehouses and storage areas. These include:

- *High-value tools for maintenance and troubleshooting.* Examples are specialized monitoring equipment such as vibration sensors, temperature sensors, air flow meters, liquid flow meters, and laptop computers with interfaces to connect with such sensors.
- *Low-value widgets such as nails and bolts.* Examples include high-volume, low-cost items.

- *Small gear and supplies.* Examples include items such as gloves, helmets, boots, etc.
- *Parts and spares for larger equipment.* Examples include pumps, impellers, valves, and fences.
- *Electrical and electronic equipment.* Examples include servers, computers, motor switches, and starters.

When a work order is issued to a service or maintenance crew to conduct a repair, the crew members visit a warehouse or storage area to obtain the material required to conduct the repair. Each work order may require items from different stock keeping units (SKUs). An example is a work order for a pump station. The crew may need two gloves, mineral oil for removing rust, a specialty wrench that is to be issued from a tool crib and returned, and process-monitoring equipment such as a flow meter and handheld computer that are to be returned after the work order is complete. The items required are collected off the shelves by the staff at the warehouse and brought to the checkout counter where they are issued to the maintenance crew. In some instances, the maintenance person may walk through a section of the warehouse and pick out the specialty equipment required for the task.

RFID technologies can be applied together with bar code scanners to improve the management of such equipment needed for work orders. The goals of the RFID-based system include the following:

- Improve the utilization of specialty equipment
- Manage the inventory better to prevent delays (from lack of availability of specialty equipment/stockouts)
- Minimize the amount of time the equipment is sitting in inventory
- Reduce the shrinkage (loss due to theft or misplacement)

Fleet. Water utilities maintain a fleet of vehicles that the work crews use to inspect and service the equipment at various facilities (such as pump stations and plants). The fleet could include cars, small to large trucks, earth moving equipment (of different sizes), cranes and forklifts, and specialty vehicles for cleaning out pipelines. In some instances, the utility may maintain additional equipment for snow removal and garbage collection.

These vehicles are parked at maintenance/vehicle yards. Once a maintenance work order enters the system, the staff assigned to the work order requests specific vehicles required to complete the work order. The vehicle fleet manager has to make these vehicles available with minimal delay.

The vehicle fleet manager has to track the usage of the vehicles. He has to balance competing demands to reduce the number of vehicles that are idle (which requires reducing the size of the fleet) and to minimize the delays due to unavailability of vehicles. The latter could require increasing the size of the fleet or insuring an access to vehicles that can be rented from other parties. In addition to tracking vehicle usage and delays due to unavailability, the manager has to track the maintenance schedule and the consumption of parts and fuel, as well as make decisions regarding types of vehicles and parts to order in the future.

The application of RFID-based technology and enterprise software allows the fleet manager to maximize the time a vehicle is utilized. This can help improve the workflow and improve the utilization and operation of vehicles and maintenance yards. The general applications are discussed in earlier chapters. Specific applications within the water utility are discussed in this chapter.

Monitoring

The water and wastewater utilities have to monitor and maintain facilities that are geographically distributed. Many of these facilities are managed remotely or staffed for only a part of the day. These facilities range in size from small pump stations to larger treatment plants.

The utility may have a control system such as a SCADA system that allows several facilities to be operated remotely. Installation of a SCADA system requires a significant investment in instrumentation, control, and automation. The SCADA systems have to be integrated with online process monitoring, especially at the medium and larger sized plants that treat more than 10,000 cubic meters per day (m^3/d) of wastewater. During emergency conditions, such as during a flood or a storm, the SCADA system can be used to prioritize staffing and maintain a certain critical level of service.

Within a SCADA system, the readings from sensors monitoring the status of equipment are converted to electronic signals and transmitted to a central location. The method of signal transmission can be hardwired or wireless. The person monitoring the SCADA system is notified of alarms. Depending on the nature of the alarm, he or she will call in personnel to service the alarm. The SCADA system is also used to troubleshoot and develop workarounds to problems (such as turning off a blower in an aeration system if it is running at a temperature higher than is normal and turning on a standby blower) from a remote location.

The presence of an IT backbone servicing a SCADA system makes it easier to incorporate RFID- and GPS-based technologies. An example is the use of a chip that can store 120 days worth of hourly flow data from a water meter and transmit the data when activated by a signal from a handheld device that is 30 m (100 ft) away. This allows the utility to not only obtain a single water meter reading, it also provides information on the pattern of usage by customers and detects leaks to prioritize repairs.

Assets in solid waste collection and public works operations

In some instances, the assets of a water utility are combined with other aspects of public works functions of a town or a city. In this case, the utility may operate garbage collection services and conduct tasks such as road maintenance. This requires operation and maintenance of additional specialty vehicles; vehicle tracking; tracking of requests for garbage collection and services by household, type, and size; and customer billing. RFID-based tags on vehicles and garbage bins can help utilities track and bill collection services by frequency, type, weight, and location.

Analysis of Problems and Solutions

This section describes five categories of problems for which solutions can be developed using RFID and wireless technologies. These five categories are presented as a matrix in table 5–1. The matrix helps relate processes at water utilities to problems faced.

Fixed assets

Description of problem—locating assets. Utilities continue to face problems with the location of aging fixed assets. These include pipelines, electrical conduits, etc., that are located underground or between floors or walls of buildings.

One recurring problem is the damage to adjacent lines that occurs during excavation to access and repair a pipeline. As mentioned previously, this is due to the proximity of lines to other structures and utility lines (fig. 5–9), their colocation within the same vertical plane, and their deviations from the locations shown in the drawings.

The challenge in developing a solution to this problem is the high level of accuracy required for locating the lines. The solution must include a method of incorporating information on the location of lines from the drawings into a database with a graphical interface, as well as an ability to update the information from additional field surveys and explorations during subsequent projects. There has to be a method for incorporating discoveries made during an exploratory survey into a database for locating assets. The information from this database has to be accessible in the field.

Table 5–1. Five categories of problems that can be addressed with RFID and wireless technologies at water utilities

Water Utility Process #	Description of Assets	Categories of Problems				
		Fixed Assets	Mobile Assets	Vehicle Tracking, Fleet Management	Secure Access	Monitoring/Recording
1. Water Supply	Dams, reservoirs, wells, river intake structures, pumps, aquaducts, pipes	Location + status of wells, location of underground pipelines	Equipment checked out for maintenance, loaded in service vehicles	Service vehicles tracking, utilization, parts & maintenance	Certain areas of dams and reservoirs	Elevation and settling of dams, structures (survey data recorded)
2. Water Treatment	Pumps, pipes, tanks, chemical feed systems, chemical precipitates, filtration, disinfection systems	Location of underground assets	Monitoring equipment + 1	Trucks hauling chemical sludge, chemicals	Certain electrical equipment and chemical storage areas	Maintenance and replacement records on valves and motors; data logging and wireless transmission of flow and water quality data.
3. Water Distribution	Pipelines, storage tanks, meters	Location of underground assets	Same as 2	Same as 1		Pressure, flow, consumption, quality, maintenance records
4. Wastewater Collection	Pipelines, pumps,		Same as 2	Same as 1		Flow, operating status of pumps
5. Wastewater Treatment	Screens, treatment tanks and equipment, pumps, pipelines, blowers, filtration, disinfection, chemical feed		Same as 2	Trucks hauling biosolids, chemicals	Same as 2	Same as 2
6. Centralized Operations	Warehouses, vehicle fleets		Same as 2 + Inventory	Same as 1		

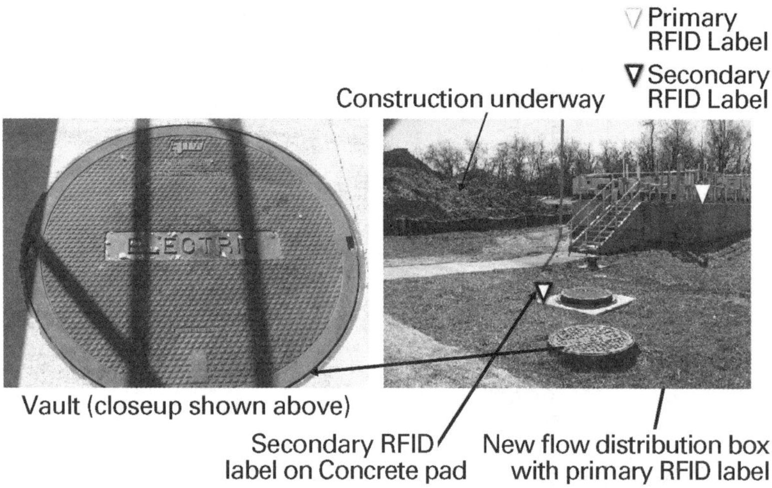

Fig. 5–9. Primary and secondary markers to label and locate electric and other buried utilities. RFID tag on the concrete pad could have information on number, locations, types of electrical conduits (fiber optic, copper, date of installation, from where to where). Note that several conduits come together at this location. Even though the plant has been expanded 4 to 5 times since the 1970s, the location of the box is unchanged. More conduits have to be run through it with each expansion. The tag is located by scanning a primary marker on the corner of the tank.

Solution for locating assets. This section discusses how solutions can be developed using an RFID-based permanent staking and marking system. One solution under consideration is the installation of markers (labels) embedded with RFID chips at the corners of buildings, on fixtures, or at locations along the edge of a tank, etc. (fig. 5–10). The coordinates of points along a line are stored relative to the locations of the primary and secondary RFID labels. The primary labels are used to locate the secondary labels. The secondary labels are located on or in the vicinity of assets. This information is accessed through a handheld device. The coordinates locate various points along each pipeline (chapter 4). The spacing between adjacent points along a pipeline stored in the database is based on several parameters. These parameters include:

- The location of adjacent ends of each straight section of pipeline.

- The locations where adjacent lines cross over each other or come close to each other.
- The number of points required to describe the curvature of the pipeline at its bends. (The number of points selected should fully describe the arc.)

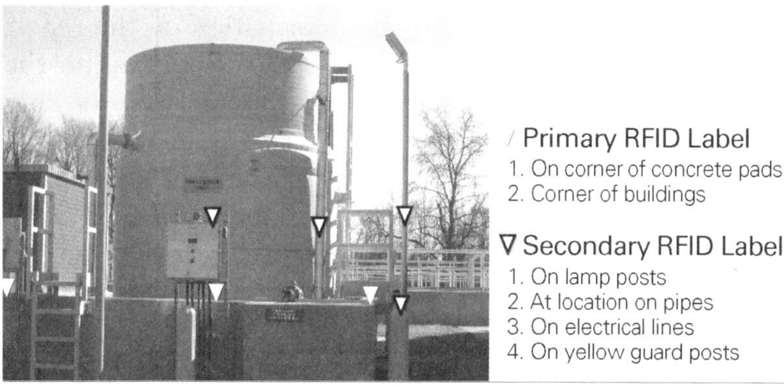

Fig. 5–10. Primary and secondary markers to label and locate fixtures for a chemical feed system

The chips selected as part of the labeling system should survive over an extended period of time (20 years). The locations of the markers should be such that they are permanent (not damaged or moved during maintenance or construction). Only when major construction is done would the markers and labels be removed and replaced by additional locations that are labeled. The labeling system must have a level of redundancy such that when a point is lost (due to storm damage, construction damage, a chip goes out of service, etc.), the location of the pipeline can be determined by the coordinates derived from other markers.

Description of problem—identification and status of assets. A second problem facing field crews is the identification of similar assets. For example, when a crew is in the field, it takes considerable effort to distinguish one valve along a pipeline from another in the vicinity. If there are parallel underground pipelines, the crew is unable to distinguish the lines.

Even after the assets have been located (e.g., valves on parallel pipes), the crew is unable to determine or verify when the last repair was conducted unless the information was entered into a database. In some instances, the record may differ from the condition observed in the field (valve appears newer or older than what is in the database).

Solution for identification and status of assets. This problem can be overcome if the information on the repair can be stored by the crew in an RFID chip placed on the asset or in the vicinity of the asset. The memory on the chip can be used to store the last repair or service date. A level of redundancy can be maintained by also storing this information on a database. The data is entered in a handheld device in the field and synchronized with the enterprise system at the end of the day. The operator can log into the database using his or her ID or an RFID-based access card. When the pipe vault is accessed in the future, the crew can obtain a confirmation of the maintenance record (such as last maintenance date) by using a handheld device to read the information stored in the memory of the RFID chip.

Mobile assets

Description of problems with tool tracking. Each water utility has a certain number of mobile assets that are critical for troubleshooting problems. Examples can include handheld computers, vibration sensors, temperature sensors, data loggers, and total stations, along with tools that are part of a tool crib, such as specialty wrenches and drills. These units are housed at a few locations (such as warehouses) within the service area of the water utility.

Once a service crew is assigned a work order to repair a piece of equipment, its members identify what tools are necessary to conduct the repair. They go to the warehouse and request the equipment. If the warehouse staff is unable to track the critical tools, it results in delays to work orders.

To make this process efficient, the warehouse staff should be capable of making following decisions:

1. Immediately determine whether all the tools necessary for the repair are available.
2. If some tools are unavailable, determine the time at which they will be returned to the warehouse and become available.
3. Determine the condition the tools are in, both at the time when they are being returned to the warehouse and when they are being reissued.
4. Determine if the tools need to be repaired before they are reissued.
5. Set aside and send damaged tools for repairs; order replacement tools.
6. Quantify the number of tools of each type that need to be in stock to prevent delays for the work crews (size of inventory required).
7. Determine the best location to store tools (which warehouse should store which tools and how many).
8. Reduce the amount of time tools are sitting unutilized (both with the service crews and on the shelf).
9. Determine techniques to reduce loss and misplacement of tools.

Developing RFID-based solutions for tool tracking. Fortunately, the needs for mobile asset management within the water utility are simpler than applications at other high-volume industries such as retail. For example:

1. There are a limited number of critical assets that need to be tracked—typically, less than

100. These consist of moderate to high-value mobile equipment.

2. The number of units of each critical asset is limited. There may be 1 to 100 units in circulation within a water utility (e.g., a handheld computer).
3. Units are housed at a few locations.

In implementing a solution, the first step is to identify those mobile assets that are critical to the operation of the utility. For example:

1. A decision can be made to label only those units that are of high value.
2. A handheld scanner can be used to read tags. This can be supplemented with an additional bar coding or detection system.
3. Scanning can be done one unit at a time because there is limited throughput. This allows opportunities to minimize interference.

This strategy has been used as part of solutions for mobile asset tracking that are in the market (e.g., Houndware by Verichip). The solutions use a combination of bar codes and RFID chips. Once the asset tracking system is in place, the software system is set up to improve the management of the assets.

Problems with fleet management. There are two different aspects to the problems with fleet management:

1. There are several types and a large number of vehicles required for different operations.
2. The number of vehicles hauling various types of chemicals and residuals (sludge, biosolids) into and out of facilities can be significant.

A medium to large water utility may maintain a fleet of 100 to 500 vehicles. These include:

1. Small vehicles—cars, small trucks
2. Four-wheel-drive vehicles used at job sites
3. Heavy equipment and large trucks—trucks, excavators, cranes, water trucks, etc.

The fleet may be located at the treatment plants or in maintenance yards distributed around the service region. Some vehicles are assigned to work crews. Others are assigned on the basis of assignment of work orders. The challenges facing water utilities include:

1. Effective utilization of vehicles (reducing the idle time).
2. Management of spare parts to improve utilization and reduce fleet maintenance costs. This includes tracking the cost and longevity of parts, optimizing their inventory, and ordering parts in a timely manner.
3. Tracking fuel consumption.
4. Scheduling service.

Developing solutions for tracking vehicles as part of a maintenance fleet. RFID-based labeling presents unique opportunities to extend the capabilities for fleet management beyond what is possible with bar codes. One can label parts or vehicles at locations that cannot be scanned with a bar code. For example, the chip may be located on the inside of the windshield of a car where it can be read by a scanner installed at the entry or exit points of a vehicle yard. An example is the RFID-based FastPass system that allows a car to pass through a tollbooth with automatic payment. The utility may use a similar system to log when a vehicle leaves and reenters a parking lot, maintenance yard, or a plant. Another application of RFID technology is the labeling of an engine or a part in the car, such as a tire hub, to log the history of maintenance and replacement of parts. It may also be a part of car keys and card readers assigned to keys (as in Speedpass by ExxonMobil or Smart keys by Nissan and other car manufacturers).

The vehicle fleet management staff can use the RFID-based system and software to accomplish the following:

1. Track a vehicle and correlate the operator or maintenance crew to the vehicle
2. Track where the vehicle is parked
3. Track maintenance and parts used on each vehicle
4. Track fuel usage by vehicle
5. Track the life cycle of parts
6. Track the effectiveness of preventative maintenance

The fleet manager can then apply the information to better manage the frequency for ordering parts. If there are multiple suppliers for parts of the fleet, the tracking system may be utilized to determine which suppliers' products are more suitable for the purposes of the utility.

Chapter 4 discussed additional applications of RFID and wireless technologies as part of GPS- and satellite-based vehicle tracking systems. These can be used by a water utility that services a large geographic area.

Problems with material tracking. Water utilities generate a variety of material that could benefit from a tracking/labeling system similar to that used for tracking hazardous wastes. For example, a water utility may be storing various batches of polymers for dewatering solids. The polymers used may have to be changed with changes in the raw water source. When a batch is taken out of service, the remaining polymer in the drum is set aside in a corner of a building, with the intention that it may be used later or disposed of at a later time. Without a tracking system, it is difficult to reuse the drum or dispose of it when it is discovered several months (or years) later. The plant operator who later finds it is unable to find the manufacturing date, the shipment log on the drum, or other information that would help with the disposal.

Another example is the tracking of small quantities of waste material, some of which may be hazardous. The quantities generated may be small but can accumulate over time. Without a continuous tracking system, it is a challenge to identify the date or dates on which the material was generated and the treatment processes that generated the material, and to determine whether it is hazardous or not. Subsequent tests to identify whether the material is hazardous are considerably more expensive.

Solutions for material tracking. The use of RFID chips to label drums and containers, however small, can provide an attractive solution for materials tracking and inventory management. Each bottle or drum can be labeled with a chip that has a unique ID number. The operator may program in a code that describes the waste into the memory of the chip and store it in a database. When the waste has to be identified prior to disposal, the operator has to scan the chip to obtain the requisite information regarding the source and nature of the waste.

Access management and infrastructure security

Most water utilities have key-based access to their facilities (i.e., they use keys rather than access cards to open doors to buildings). This is a reliable means of access that has proven itself over the years. The limitations with key-based access are as follows:

1. There is no log describing who accessed a restricted facility at what time.
2. For facilities that need routine inspection, there is no separate log of the inspection.
3. There is no log of the vehicles that entered facility and the types of wastes brought in or carried out of a facility.

Installing a card reader can help address several of these limitations, such as the following:

1. Maintaining a log of access (important for passing inspection audits)
2. Ability to integrate information of reading on an instrument to operator
3. Controlling access, if required

An RFID-based solution can be implemented as an alternative to card readers. For example, a utility may implement the following to maintain a record of wastes hauled into or out of a facility:

1. Install RFID chips on all vehicles that may access a facility (on fleets used by the utility and by the haulers).
2. Automatically record which vehicle accessed the facility.
3. Require the vehicle operator to scan a badge or enter a number to open a valve and let the waste flow into a tank (e.g., an industrial waste hauler or a septage waste hauler bringing in material to a plant). The quantity of the waste flow is recorded by a meter in the tank.

The system can be expanded to track vehicles that remove solid wastes (sludge) from a plant. A similar system can be put into place for cities and utilities that have collection points for recycled material (oil, batteries, etc.).

Monitoring and recording

Challenges in monitoring and recording. Most water utilities in the United States face more stringent needs for monitoring and recording data, either because of changes in regulatory requirements or because of the need to remotely operate facilities and reduce operating costs. In recent years, water utilities have been able to leverage monitoring requirements for regulatory compliance together with the monitoring needs to remotely operate plants and pump stations. One example is the data from the monitoring of treated water or wastewater quality from a treatment plant. This has to be reported for regulatory compliance. This data on the treated water can also be used to determine if the upstream treatment processes are operating effectively. The deviations from the norm are detected using the monitoring equipment that is operating continuously, rather than additional sampling and laboratory analysis of the samples a few days later. The plant operator is able to respond in a timely manner and correct the problem. While the cost of monitoring continues to be a challenge, recent advances in RFID and nanotechnologies are helping develop nanoprobes and RFID-based data transmission devices that may help reduce this cost for some applications.

Solutions for monitoring. The integration of RFID chips with other metering and data recording systems presents some unique opportunities to increase monitoring and lower costs. RFID chips can be integrated with measuring devices (such as temperature sensors, vibration sensors, or flow meters) and data loggers (which store several measurements in their memory or on the chip). Several hours' data can be transmitted wirelessly to a computer. The chip can be activated remotely and instructed to transmit data. This capability is being used by water utilities to obtain hourly records of water consumption data from water meters.

An example is a solution from the Hervey Bay area of Queensland, Australia (Wide Bay Water Corp.) that uses three pieces of equipment:

- An Elster meter that reads the water flow
- A Datamatic data logger (Firefly) that can store and transmit 120 days worth of hourly data using a wireless frequency to transmit the data

- A Datamatic transceiver called *Road Runner* that is used to remotely activate the data logger and download the data.[1]

Solutions for data recording. RFID chips can be used to label assets in a manner that improves the management of the asset. The memory of the chip may contain data that serves as a unique identifier. In this instance, a handheld computer has to use this identifier to access a database that contains a history of maintenance on the asset. The database may reside on an enterprise system. The latest data from the database may be downloaded to the handheld device. The same functionality can be incorporated using a bar code except for the fact that the RFID chip does not require direct line of sight. The asset would still be tagged with a unique ID in case the chip fails.

In a more advanced application, the memory chip could be used to program the last service date, in addition to its unique identifier. This frees up the need to access a database.

The types of data records that may benefit from such a system include the following:

1. Finding a record of data from earlier surveys (elevation of dam, location of line)
2. A record of date and type of activity for maintenance performed on an asset
    - Repair of equipment
    - Oil change
    - Weight of trash removed from a tank
    - Weight of grit, screenings, or biosolids hauled by each truck and date
    - Nature and type of garbage or recyclable material removed from each house, plant, or facility by date

The system may be designed to provide better access to the information in the field to help with decision making.

Costs and Benefits and Improving the Return on Invested Capital

The most desirable outcome of an RFID and wireless technology implementation is a low-cost, high-benefit, large return on the invested capital. For this, the authors recommend that the utility conduct an asset management plan through which it identifies the most critical problems. These are problems that disrupt the normal functioning of the utility, pose significant risk to personnel, or result in a substantial cost or liability. Once the problems are prioritized, the next step is to move to implementation.

The implementation of RFID- and wireless-based solutions requires an understanding of the hardware and software discussed in previous chapters. It is recommended that the solutions to the five categories of problems as discussed be implemented as separate or as phased/sequential projects. Certain projects such as those for monitoring can be implemented as part of plant-wide instrumentation projects. Others such as mobile asset management can be implemented within a warehouse. The vehicle management can be initiated at vehicle maintenance, storage, and refueling areas. The static asset management can be implemented as part of a plant expansion.

Appendix A contains a list of costs and benefits that go into an ROI computation.

The authors would like to acknowledge the assistance of Greg Farren who managed the operations of the Annapolis WRF during its upgrades from 1991 to 2006.

References

1. International Water Association. 2007. Water metering: swing to smart solutions. *Water21*. IWA Publishing. April.

2. Annapolis Water Reclamation Facility. 2006. Anne Arundel County, MD.

3. American Water Works Association (AWWA). 2008. Drinking Water Week. www.awwa.org.

6 RFID AND WIRELESS APPLICATIONS FOR ASSET MANAGEMENT IN THE ELECTRIC UTILITY

Introduction

The electric utility industry provides electrical power (energy) to households and businesses. The industry is split into three processes—generation, transmission, and distribution (fig. 6–1). In some instances, the distribution may be separated further into distribution and retail.

The ownership of the electrical infrastructure differs by geography. In some locations, the entire infrastructure is owned by a government entity or a public entity that has to get electric rates approved by a commission. In other areas, the industry has been deregulated and different parts of the infrastructure have been privatized.

Each entity involved in the electric utility industry manages assets that are geographically distributed within the service area. The assets may be fixed or mobile. There are many areas in which RFID-based solutions can increase the efficiency of the business processes. Applications could include the tracking of assets for replacement, utilization, and loss; automated monitoring of usage and losses along the lines; maintenance of a fleet of vehicles; and automated metering and demand management.

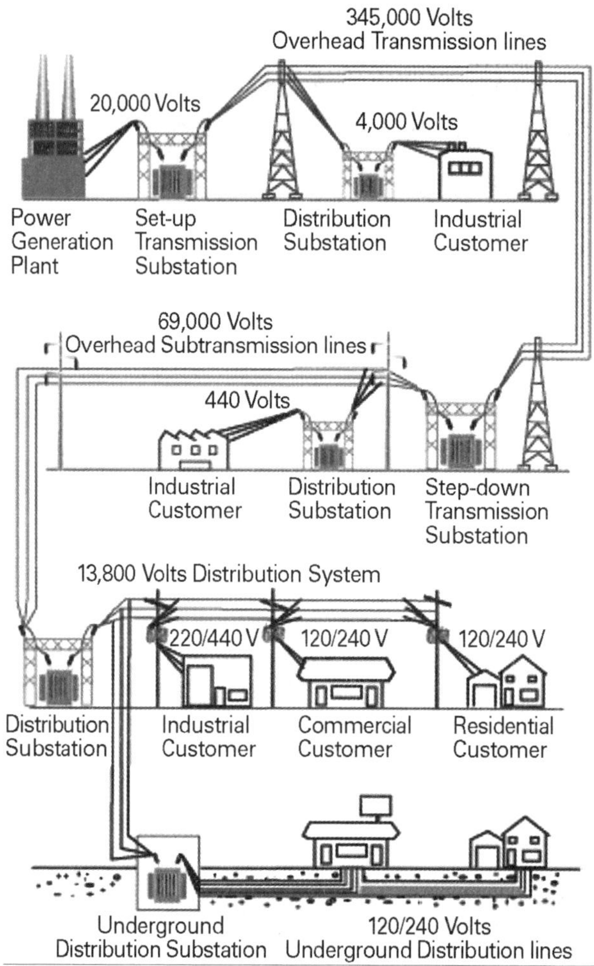

Fig. 6–1. Schematic of the power industry showing generation, transmission, and distribution. *Source: OSHA, 2007*

Description of Infrastructure

Power generation

Much of the electricity generated today is from rotating turbines attached to electrical generators. This significant infrastructure can occupy a large land area that comprises the power plant. There is considerable opportunity to improve the tracking and management of assets with the use of RFID within the power plant.

Since much of the electrical generation today is with coal-fired power plants, it makes sense to review a schematic of such a plant (fig. 6–2) to understand the array of assets within the plant (fig. 6–3). Gas-fired plants have several components that are similar.

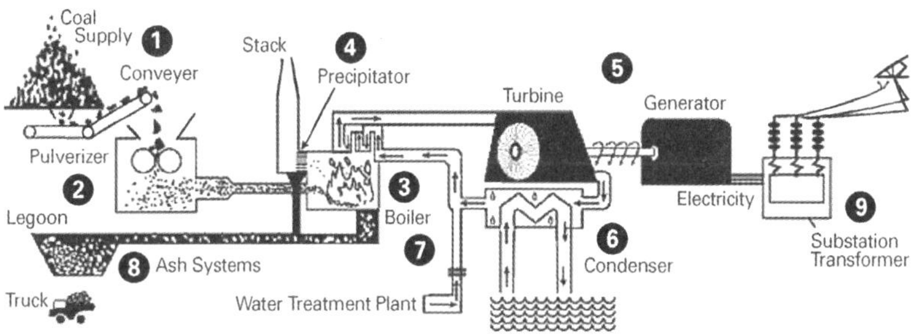

Fig. 6–2. Schematic of a coal-fired thermal power plant. *Source: Canadian Clean Power Coalition*

Fig. 6–3. Singrauli and Bokaro Power Plants, India. *Source: Ministry of Power, India, 2004 and NTPC, 2006*

Coal supply and consumption

The coal is transported by rail cars or barged to a site within or adjacent to the power plant. Sufficient coal is stored at the site to get over periods when it may be difficult to supply the plant, such as in winter when it may be difficult to work under a blanket of snow and ice. Plants have developed methods of keeping track of supply and the inventory of coal on hand. Application of RFID on rail cars and barges may be one way of automating the tracking of the supply of coal. The consumption can be monitored by monitoring the hours of operation of the coal conveyor (fig. 6–2, item 1) and the weight of coal being conveyed per hour. Such monitoring devices and transponders can be radio frequency enabled.

Boiler and associated equipment

The coal conveyed by the conveyor is pulverized to a fine powder to increase the surface area of coal and its exposure to preheated air. This helps create a heated air and fuel mixture that is sent at a high pressure into the blower, where it ignites. The boiler is a furnace that has a height of 40 m (130 ft) and a base of 15 m x 15 m (50 ft x 50 ft). The walls are made of high-pressure steel tubes. Its purpose is to produce steam at the high purity, pressure, and temperature required for the steam turbine. The turbine drives the generator.

As water with a high purity passes up the tubes on the walls of the boiler (item 3), it turns into steam. The steam is separated from the water in a boiler drum. The steam is conveyed to a superheater, where its temperature and pressure are increased sufficiently to run a high-pressure turbine.

The exhaust gases from the boiler are suctioned by a draft fan through an electrostatic precipitator (item 4) to an exhaust gas (chimney) stack. Additional devices may be installed for removal of sulfur dioxide, nitrous oxides, and particulates.

Turbine

The steam turbine (item 5) is comprised of high-pressure, intermediate-pressure, and low-pressure sets on a single shaft. The shaft is coupled to the three-phase electrical generator (item 5). The shaft is heavy and has to be kept in position while running with bearings. The bearings are lined with low-friction material and lubricated to minimize the frictional resistance. Barring gear is used to keep the shaft rotating at a very low speed when the unit has tripped (stopped). This rotation reduces the risk of the turbine shaft warping as it cools down.

Condenser

Steam exiting the high-pressure turbine is passed through a boiler reheater before it is sent to the intermediate pressure turbine. It is then conveyed to the low-pressure turbine. The steam leaving the low-pressure turbine is above the boiling point of water. It is brought into indirect contact with cooling water from the cooling tower in a condenser (item 6) to condense it. The condenser is a shell and tube heat exchanger (fig. 6–4) in which cooling water is circulated through the tubes. The condensed water is pumped to a deaerator (12). The cooling water is cooled by spraying it inside a cooling tower. Makeup cooling water is taken directly from a river, lake, or ocean located in the vicinity of the power plant. In a few instances, it is from treated wastewater discharged from a wastewater treatment plant (such as the Mountain View Gas-fired Power Plant, Redlands, CA, which receives water treated through a membrane bioreactor at Redlands wastewater treatment facility).

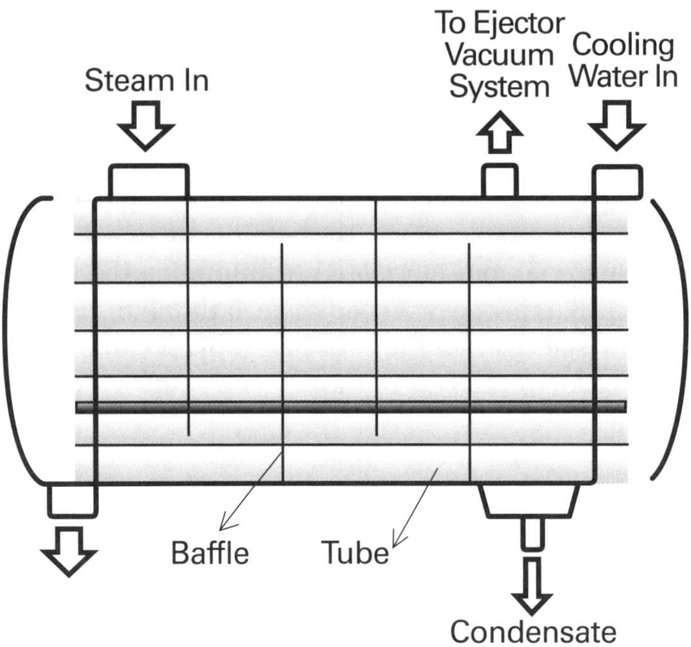

Fig. 6–4. Schematic of a shell and tube condenser. Cooling water enters from the right side of this single-pass condenser, flows through the tubes, and exits on the left side. The steam passes along the outside of the tubes and is routed by the baffles. The condensate that is formed leaves at the bottom of the vessel. The remaining vapor and air is removed by the ejector vacuum system.

The condensate (fig 6–4) is warmed in the feed heater and sent to the boiler feed pump after passing it through a deaerator (fig. 6–5). The deaerator is discussed in the following section. The condensate water is prewarmed, first in a feed heater powered by steam drawn from the high-pressure set, and then in an economizer, before being returned to the boiler drum. The cooling water from the condenser is sprayed inside a cooling tower, creating a highly visible plume of water vapor, before being pumped back to the condenser in the cooling water cycle.

Boiler feed water deaerator

A steam-generating boiler requires water that is devoid of dissolved gases to avoid corrosion to the metal. Power stations use a deaerator for this purpose (fig. 6–5). A deaerator includes a vertical-domed deaeration section with trays where the boiler feed water is deaerated using steam. The deaerated liquid is stored and heated (with steam) in a storage reservoir below. The waste gases are vented at the top.

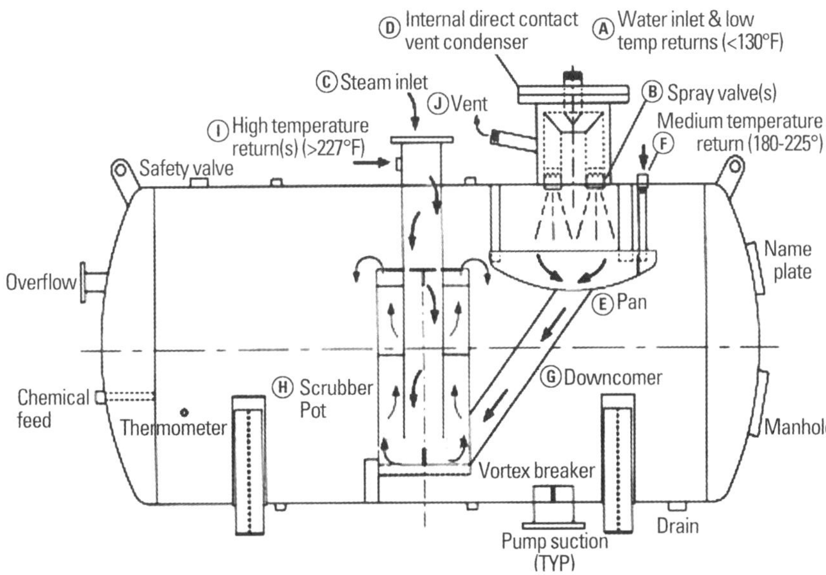

Fig. 6–5. Schematic of a deaerator used to remove air from water used as boiler feed water. *Source: Pascual and LeMay, 2007.*

Generator

The turbines are coupled on the same shaft to a three-phase generator (fig. 6–2, item 5). The generator voltage varies from 11 KV in smaller units to 22 KV in larger units. The high-voltage leads are large aluminum channels to convey the high current. The leads are connected to step up transformers that increase the voltage to 110 KV or 220 KV for transmission to a power grid.

The generator has a cooling system to dissipate the heat. Small units may be air cooled. Larger units may be cooled with water or hydrogen gas.

Exhaust gas treatment system

Fly ash is present in the gas coming off the boilers. Fly ash is comprised of tiny particles of ash. Fabric bag filters or electrostatic precipitators (fig. 6–2, item 4) are used to remove the fly ash. The gases leaving the electrostatic precipitators contain oxides of nitrogen and sulfur. These can be removed in gas scrubbers. The scrubbers use a crushed limestone bed, slurries, and alkaline spray mists to treat the gas.

The gas that remains is discharged by the stack at a certain height above the ground. This height is selected by atmospheric dispersion modeling to insure good dispersion of the fine particulates and acidic gases that remain after treatment.

Power transmission

Electric power transmission systems are used to transfer the power from a power plant to an electrical substation located in the vicinity of a residential or business area. Multiple and distant power plants can be located on the network that is termed as a *grid*. The transmission is at a high voltage (110 KV and higher) through overhead power lines (fig. 6–6).

In densely populated urban areas, the transmission may be through underground power lines. There are considerable additional costs associated with cooling, reactive power, and voltage management when high-voltage transmission lines are underground.

Fig. 6–6. Transmission towers and lines in Southern California

The transmission lines are made of several strands of aluminum alloy reinforced with strands of steel. The wires are suspended off the towers using insulators.

The transmission grid is connected to substations at different points along the grid (figs. 6–1 and 6–7). The substation decreases the voltage from the high-voltage grid to a local grid for distribution. In some instances, the substation may be a booster station to allow the power to travel greater distances.

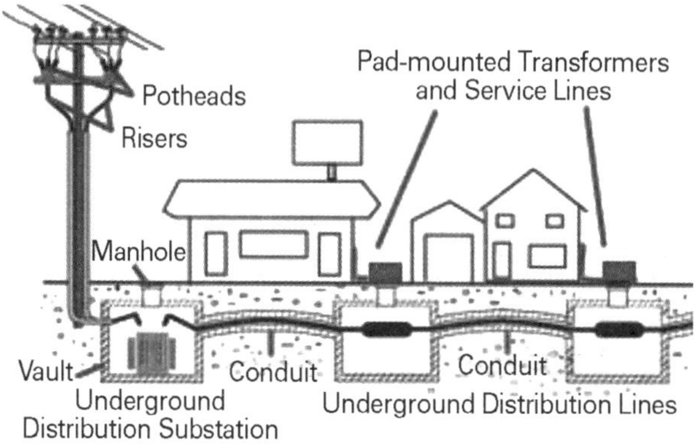

Fig. 6–7. Schematic of secondary distribution system

There is considerable opportunity to apply radio frequency and RFID devices to improve the tracking, monitoring, and management of power transmission systems. These can include simple tasks such as labeling transformers and towers with RFID chips to monitoring devices equipped with radio transmission technology. The memory in the RFID chips can store the unique serial number. This can be a backup to a painted serial number. The ID of the chip can be read by a handheld device and correlated to information regarding its maintenance. The memory in the chip can also be used to store some critical information that can be downloaded locally (temperature, last maintenance date, etc.).

Power distribution and retail

The power distribution and retail systems include the infrastructure to carry electricity from the high-voltage transmission systems to the end customer (fig. 6–1). The assets for power distribution include a series of step down subtransmission and primary distribution systems (fig. 6–1). This is followed by a secondary distribution system that includes the poles, transformers, reclosers, capacitors, voltage regulators, and electric meters (fig. 6–7). The geographic distribution and the high numbers for each asset necessitate the use of low-cost standardized items. The items should be reliable (have life spans of 20 or more years) and easy to replace.

As power begins to exit the transmission system, the voltage is reduced from 220 KV or higher voltage to 69 to 169 KV for conveyance through the subtransmission systems. It is then reduced to 4 to 35 KV in substations located in towns or neighborhoods and fed to the primary distribution system (figs. 6–8 and 6–9). The primary distribution system feeds the poles running along the streets (seen in older neighborhoods) or underground lines.

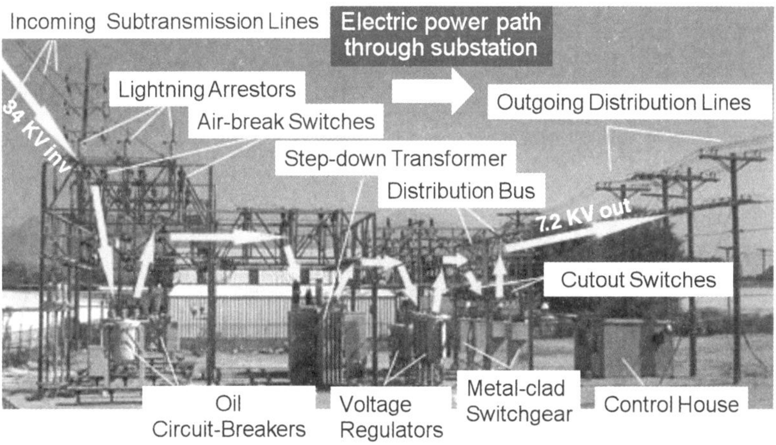

Fig. 6–8. Assets within a substation to step down voltage from subtransmission system to primary distribution system. *Source: OSHA, 2008*

Fig. 6–9. Electrical substation, Northern California. Picture of a PGE substation, Merced, California

Transformers can be located on the poles (figs. 6–10 and 6–11) or in underground vaults (fig. 6–12). These transformers reduce the voltage to 120 V in the U.S. markets and 240 V in European markets and connect to the secondary distribution system. The poles or vaults are located 50 to 300 ft from homes.

Fig. 6–10. Light pole with transformer to reduce voltage to 120 V for distribution

Fig. 6–11. Close-up of light pole with transformer

Fig. 6–12. Underground vault housing with step down transformer

Typically, the utility labels each pole with a unique serial number. Some of these poles may have a street light; others may have a transformer and a street light (fig. 6–10). Each underground vault housing a transformer may also have a unique serial number.

During maintenance, the operators have to identify the serial number on the pole or the vault and correlate it to the type of fixtures associated with it. The application of RFID chips on the pole or vault may allow the field crew to store information on the serial number and the type of fixtures on the read/write memory of the chip. It may also be able to store information on the houses served. During a service call, the crew should be able to access this information through a handheld device. It will help the crew confirm the types of fixtures, identify the homes served by the fixtures, and confirm the maintenance history in the field. Overall, this can improve the reliability of the system (including repairs completed) and reduce the amount of time taken to perform the service.

Analysis of Problems and Solutions

This section describes five categories of problems for which solutions can be developed using RFID and wireless technologies. These five categories are presented as a matrix in table 6–1. The matrix helps relate the processes at electric utilities to the problems faced.

Fixed assets

Description of problems—locating assets. Utilities that provide gas and electric service to homes continue to face problems with locating aging fixed assets. These include pipelines, electrical conduits, etc. that are located underground or between floors or walls of buildings.

Table 6–1. Categories of problems that can be addressed with RFID and wireless technologies at electric utilities

Electric Utility Process #	Description of Assets	Categories of Problems				
		Fixed Assets	Mobile Assets	Vehicle Tracking, Fleet Management	Secure Access	Monitoring/Recording
1. Hydro power generation	Dams, reservoirs, wells, river intake structures, pumps, aquaducts, pipes	Location + status of wells, location of underground pipelines	Equipment checked out for maintenance, loaded in service vehicles	Service vehicles tracking, utilization, parts & maintenance	Certain areas of dams and reservoirs	Elevation and settling of dams, structures (survey data recorded)
2. Power Plant		Location of underground and hidden assets	Monitoring equipment + 1	Rail cars, Trucks hauling coal, chemicals, fuel, gas + 1	Certain equipment, fuel and storage areas	Maintenance and replacement records on equipment; data logging and wireless transmission of emissions, temperature, vibration, voltage, etc.
3. Electric Transmission	Power Lines, towers	Location, history, maintenance, clearing, washing	Same as 2	Same as 1		Temperature, losses, dates of clearing of vegetation, washing, monitoring condition of lines
4. Electric Distribution and Retail	Transformers, electrical lines, switchers, reclosers, insulators, vaults	Location of assets, maintenance history	Same as 2	Same as 1		Same as 3
5. Centralized Operations	Warehouses, vehicle fleets		Same as 2 + Inventory	Same as 1		

One recurring problem is the damage to adjacent lines that can occur during excavation to access and repair a power line or a gas line. This is because of the proximity of lines to other structures and utility lines (fig. 6–13), their colocation within the same vertical plane, and their deviations from the locations shown in the drawings.

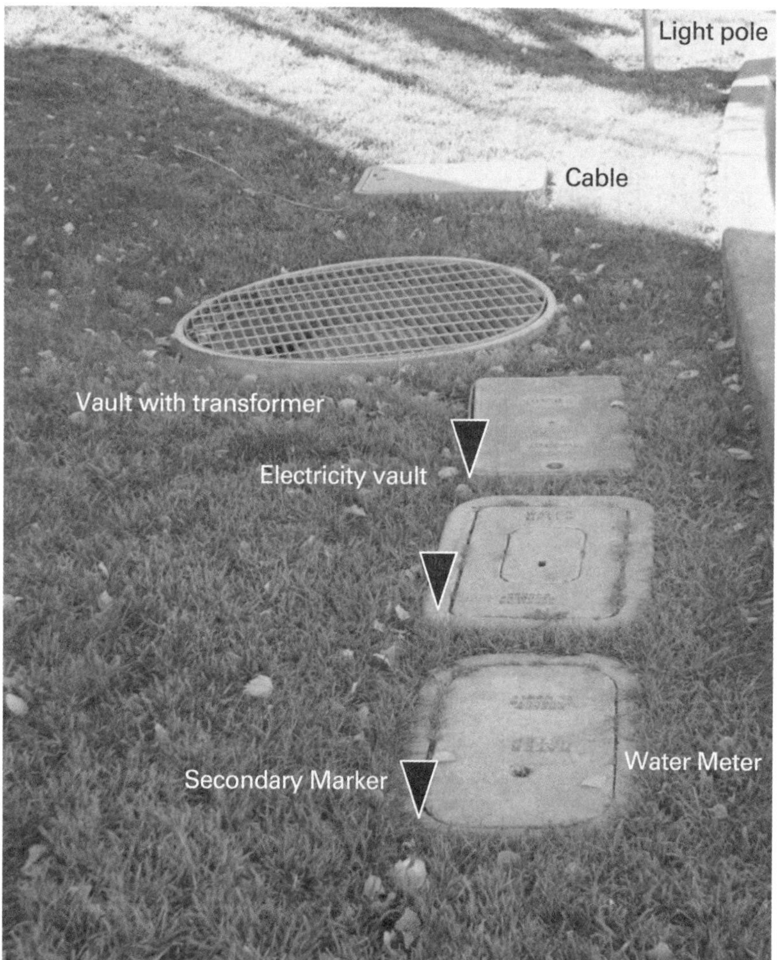

Fig. 6–13. Proximity of cable, water, and electrical utilities

Solution for locating assets. This section discusses how a solution can be developed using an RFID-based permanent staking and marking system. One solution under consideration is the installation of markers (labels) embedded with RFID chips at light poles, along the corners of buildings, on fixtures, or at locations along the edge of a substation or a vault (figs. 6–14 and 6–15). The coordinates of points along a line are stored relative to the locations of the primary and secondary RFID labels. The primary labels are used to locate the secondary labels. The secondary labels are located on or in the vicinity of the assets. This information is accessed through a handheld device. The coordinates locate various points along each conduit or pipeline (chapter 4). The spacing between adjacent points along a conduit or pipeline stored in the database is based on several parameters, such as the following:

- The location of adjacent ends of each straight section of pipeline.
- Locations where adjacent lines cross over each other or come close to each other.
- The number of points required to describe the curvature of the pipeline at its bends. (The number of points selected should fully describe the arc of the bend.)

Fig. 6–14. Use of primary labels (markers) on light pole and curb to identify location of secondary markers. The secondary markers located on individual vaults identify the nature of the utility and the homes that are connected through the vaults.

Fig. 6–15. The primary label (marker) is an RFID chip located adjacent to the number on the pole.

The chips selected as part of the labeling system should survive over an extended period of time (20 years). The locations of the markers should be such that they are permanent (not damaged or moved during maintenance or construction). The labeling system must have a level of redundancy such that when a point is lost (due to storm damage, construction damage, a chip goes out of service, etc.), the location of the conduit or pipeline can still be determined by the coordinates derived from other markers.

Description of problem—identification and status of assets. A second problem facing field crews is the identification of similar assets. For example, when a crew is in the field, it takes considerable effort to distinguish electrical wiring and service to one underground line from another in the vicinity. If there are parallel underground lines, the crew spends time distinguishing the lines.

Even after the assets have been located (e.g., parallel electrical lines and optical cables), the crew is unable to determine or verify when the last repair was conducted unless the information was entered into a database. In some instances, the record may differ from the condition observed in the field (the condition of a switch or circuit breaker appears older than what is recorded in the database).

Solution for identification and status of assets. This problem can be overcome if information on each repair can also be stored locally on the RFID chip placed on the asset or in the vicinity of the asset. The memory on the chip can be used to store the last repair or service date. A level of redundancy is maintained by also storing this information on a database. The data is entered in the field immediately after the repair using a handheld device that can read and write to the RFID chip. The operator can log into the database using his or her ID or an RFID-based access card. When the electric vault is accessed in the future, the crew can obtain a confirmation of the maintenance record (such as the last maintenance date) by using a handheld device to read the information stored in the memory of the RFID chip.

Mobile assets

Description of problems with tool tracking. Each electric utility has a certain number of mobile assets that are critical for troubleshooting problems. Examples can include handheld computers, meters for monitoring power, voltage, and current, vibration sensors, and temperature sensors. These units are housed at a few locations (such as warehouses) within the service area of the electric utility.

Once a service crew is assigned a work order to repair a piece of equipment, its members identify what tools are necessary to conduct the repair. They go to the warehouse and request the equipment. If the warehouse staff is unable to issue the tools, it results in delays to work orders.

To make this process efficient, the warehouse staff should be capable of making following decisions:

1. Immediately determine whether all the tools necessary for the repair are available.
2. If some tools are unavailable, determine the time at which they will be returned to the warehouse and can be reissued.
3. Determine the condition of the tools, both at the time when they are being returned and when they are being reissued.
4. Determine if a certain tool needs to be repaired before it can be reissued.
5. Set aside and send damaged tools for repair or order replacement tools.
6. Quantify the number of tools of each type that need to be in stock to prevent delays for the work crews (size of inventory required).
7. Determine the best location to store tools (which warehouse should store which tools and how many).
8. Reduce the amount of time tools are sitting unutilized (both with the service crews and on the shelf).
9. Determine techniques to reduce loss and misplacement of tools.

Developing RFID-based solutions for tool tracking. Fortunately, mobile asset management within the electric utility is simpler than its applications at other high-volume industries such as retail. For example:

1. One can sort through the types of equipment handled to determine the critical assets that need to be tracked. Typically, this number will

be less than 100. These consist of moderate- to high-value mobile equipment.

2. The number of units of each critical asset owned by the utility is limited. There may be 1 to 100 units in circulation within an electric utility (e.g., a handheld computer).
3. Units are housed at a few locations.

In implementing a solution, the first step is to identify those mobile assets that are critical to the operation of the utility. For example,

1. A decision can be made to label only those units that are of high value.
2. A handheld scanner can be used to read tags. This can be supplemented with an additional bar coding or detection system.
3. Scanning can be done one unit at a time because there is limited throughput. This allows opportunities to minimize interference.

This strategy has been used as part of solutions for mobile asset tracking that are in the market (e.g., Houndware by Verichip). The solutions use a combination of bar codes and RFID chips. Once the asset tracking system is in place, the software system is set up to improve the management of the assets.

Problems with fleet management. A medium to large electric utility may maintain a fleet of around 100 to 3,000 vehicles. These include:

1. Small vehicles—small trucks and cars
2. Four-wheel-drive vehicles used at job sites
3. Trucks with equipment to repair local electric poles and lines
4. Heavy equipment and large trucks—trucks, trucks with baskets, excavators, cranes, etc.

The fleet is located at maintenance yards distributed around the service region. Some vehicles are assigned to work crews. Others are assigned on the basis of work orders. The challenges facing electric utilities include the following:

1. Effective utilization of vehicles (reducing the idle time).
2. Management of spare parts to improve utilization and reduce fleet maintenance costs. This includes tracking the cost and longevity of parts, optimizing their inventory, and ordering parts in a timely manner.
3. Tracking fuel consumption.
4. Scheduling service vehicles to avoid delays.

Developing solutions for tracking vehicles as part of a maintenance fleet. RFID-based labeling presents unique opportunities that extend the capabilities for fleet management beyond what is possible with bar codes. One can label parts or vehicles at locations (on the vehicle) that are not able to be scanned with a bar code. For example, the chip may be located on the inside of the windshield of a car. Scanners installed on the gates to enter and exit the yard can read the chip. An example of RFID chips as scanners in the transportation industry is the RFID-based FastPass system that allows a car to pass through a toll booth with automatic payment. The utility uses the same system to log when a vehicle leaves and reenters the maintenance yard.

Additionally, RFID chips may be located on an engine part or on a tire hub, etc., which is not within the line of sight of a bar code. This is used to track the replacement or service dates. Additionally, the chips may be a part of car keys and card readers assigned to keys (as in Speedpass by ExxonMobil, Smart Keys by Nissan and other car manufacturers, chapter 1). The key is used to control access to a vehicle and log its fuel consumption at gas pumps.

Therefore, the vehicle fleet management staff can use the RFID-based system and software to accomplish the following:

1. Track a vehicle and correlate the operator or maintenance crew to the vehicle.
2. Track where the vehicle is parked.
3. Track maintenance and parts used on each vehicle.
4. Track fuel usage by vehicle.
5. Track the life cycle of parts.
6. Track the effectiveness of preventative maintenance.

The fleet manager can then apply the information to better manage the frequency for ordering parts. If there are multiple suppliers for parts of the fleet, the tracking system may be utilized to determine which suppliers' products are more suitable for the purposes of the utility.

Chapter 3 discusses additional applications of RFID and wireless technologies as part of GPS- and satellite-based vehicle tracking systems. These can be used by an electric utility that services a large geographic area.

Problems with material tracking. Electric utilities generate a variety of materials that could benefit from a tracking/labeling system similar to that used for tracking hazardous wastes. Small quantities of waste material (coolant oil from transformers, etc.) may be generated and can accumulate over time. Without a continuous tracking system, it is a challenge to identify the date or dates on which the material was generated and the treatment processes that generated the material, and to determine whether it is hazardous or not. Subsequent tests to identify whether the material is hazardous are considerably more expensive.

Solutions for material tracking. The use of RFID chips to label drums and containers, however small, can provide an attractive solution for materials tracking and inventory management. Each bottle or drum can be labeled with a chip that has a unique ID number. The operator may program in a code that describes the waste into the memory of the chip and stores it in a database. When the waste has to be identified prior to disposal, the operator has to scan the chip to obtain the requisite information regarding the source and nature of the waste.

Access management and infrastructure security. Most electric utilities have key-based access to their facilities (i.e., they use keys rather than access cards to open doors to buildings). This is a reliable means of access that has proven itself over the years. The limitations with key-based access are as follows:

1. There is no log describing who accessed a restricted facility at what time.
2. For facilities that need routine inspection, there is no separate log of the inspection.
3. There is no log of the vehicles that entered the facility and the types of wastes brought in or carried out of the facility.

Installing a card reader can help address several of these limitations, such as the following:

1. Maintaining a log of access (important for passing inspection audits)
2. Ability to integrate information of reading on an instrument to operator
3. Controlling access, if required

Monitoring and recording

Challenges in monitoring and recording. Some of the operations at electric utilities face more stringent needs for monitoring and recording data, either because of changes in regulatory requirements or because of the need to remotely operate facilities and reduce operating costs. These include emissions monitoring of chimney stacks and the application of SCADA to monitor substations. The data from continuous monitoring can also be used for determining if the upstream processes (such as the boiler at a coal-fired power plant) are operating effectively. When there is a deviation, the operator is able to respond in a timely manner and correct the problem.

Solutions for monitoring. The integration of RFID chips with other metering and data recording systems presents some unique opportunities to increase monitoring and lower costs. These features are part of automated meter readers being introduced into the market. RFID chips can be integrated with measuring devices (such as electric meters, temperature sensors, or vibration sensors) and data loggers (that store several consecutive measurements made over time in their memories or on the chips). The data is transmitted over wireless networks to a computer. The chip can be activated remotely and instructed to transmit data. This capability is being used by electric utilities to download records of power consumption and to dynamically manage the power consumption.

Solutions for data recording. RFID chips can be used to label assets in a manner that improves the management of the asset. The memory of the chip may contain data that serves as a unique identifier. In this instance, a handheld computer has to use this identifier to access a database that contains a history of maintenance on the asset. The database may reside on an enterprise system. The latest data from the database may be downloaded to the handheld device. The same functionality can be incorporated using a bar code except for the fact that the RFID chip does not require direct line of sight. The asset would still be tagged with a unique ID in case the chip fails.

In a more advanced application, the memory chip could be used to program the last service date, in addition to its unique identifier. This frees up the need to access a database.

The types of data records that may benefit from such a system include the following:

1. Finding a record of data from earlier surveys (elevation of dam, location of line)
2. A record of date and type of activity for maintenance performed on an asset such as repair or replacement of parts

The system may be designed to provide better access to the information in the field to help with the decision making.

Costs and Benefits and Improving the Return on Invested Capital

The most desirable outcome of an RFID and wireless technology implementation is a low-cost, high-benefit, large return on the invested capital. For this, the authors recommend that the utility conduct an asset management plan through which it identifies the most critical problems. These are problems that disrupt the normal functioning of the utility, pose significant risk to personnel, or result in a substantial cost or liability. Once the problems are prioritized, the next step is to move to implementation.

The implementation of RFID- and wireless-based solutions requires an understanding of the hardware and software discussed in previous chapters. It is recommended that the solutions to the five categories of problems as discussed be implemented as separate or phased/sequential projects. Certain projects such as those for monitoring can be implemented as part of plant-wide instrumentation projects. Others such as mobile asset management can be implemented within a warehouse. The vehicle management can be initiated at vehicle maintenance, storage, and refueling areas. The static asset management can be implemented as part of a plant expansion.

Appendix A contains a list of costs and benefits that go into an ROI computation.

References

1. "How a coal fired power plant works." 2008. Canadiancleanpowercoalition.com.
2. Pascuel, R and LeMay, I. "Deaerators: A Continuing Threat to Power Plant Safety?" 2007. 11th IPEIA Conf, Banff, Calgary, Alberta, Canada. Feb 14–17th.
3. Ministry of Power, India, 2004.
4. NTPC: National Thermal Power Corporation, India. 2006.
5. OSHA.gov. 2007.

7 RFID AND WIRELESS APPLICATIONS FOR ASSET MANAGEMENT IN THE NATURAL GAS INDUSTRY

Introduction

The natural gas industry supplies the natural gas extracted from oil and gas fields to the end customers. Several business entities are involved in the extraction, processing, transmission, and distribution functions along the natural gas supply chain. The entities may be in widely different geographies, with considerable distance between the extraction fields and the end customer. The nature of the operations and their geographic distribution present considerable opportunities to improve asset management using RFID- and wireless-based technologies. This chapter discusses how the manager for each function can consider such technologies.

The different functions and their geographic distribution are best represented using a schematic as shown in figure 7–1. The first step is the extraction of gas from the gas fields. Following extraction, the gas is processed to remove impurities. It is compressed and pressurized for transmission through major pipelines, diverted for storage in underground caverns or surface tanks, and transmitted to the local distribution companies (LDCs). In some instances, the gas is transmitted as liquefied natural gas (LNG) via rail cars or ships (tankers).

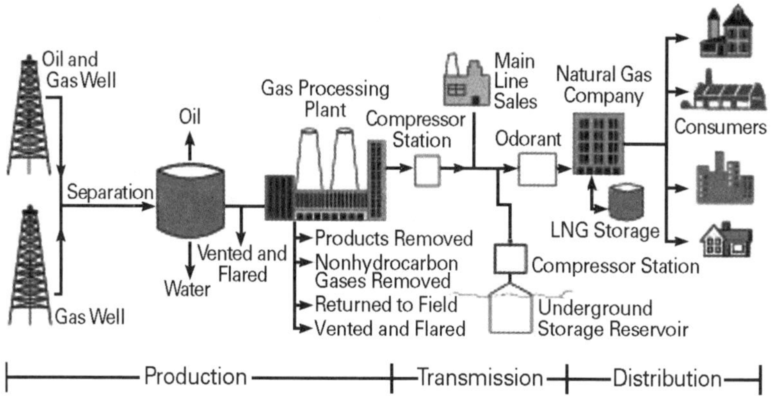

Fig. 7–1. Schematic showing the infrastructure in the natural gas industry

The LDCs reduce the pressure of the gas for distribution through local gas lines to the end customers. They add mercaptan to the gas to provide it with its rotten egg odor before distributing and metering it to the end customers. The end customers may be households, industries, natural gas retail stations, or power plants.

The exploration, extraction, and processing are pursued by large publicly traded and small privately held companies. The gas transmission lines are owned and operated by government entities or private companies. The local distribution systems are owned and managed by gas utilities, cities, and private retailers. In several instances, the electric utility is an end consumer served off the transmission or distribution system.

The natural gas distributed through the gas lines is principally comprised of methane. Heavier hydrocarbons such as propane and butane, if present, are separated at the processing plant as natural gas liquids (NGLs). These hydrocarbons are stored and distributed by additional retailers.

Description of Infrastructure

Production

Gas exploration. Gas exploration is pursued by specialty companies that identify tracts of land for exploration. These companies may own or lease the exploratory equipment. This includes special trucks with equipment to induce seismic vibrations (all-terrain seismic vibrator trucks) and geophones that are located at different points around the site to record reflections of the seismic waves generated by the trucks. Specialized equipment also includes trucks equipped with seismic recording equipment to record and analyze the data from the geophones. The data recorded is analyzed by a team of geologists and engineers to determine locations where exploratory wells should be drilled. A schematic of onshore seismic exploration is shown in figure 7–2.

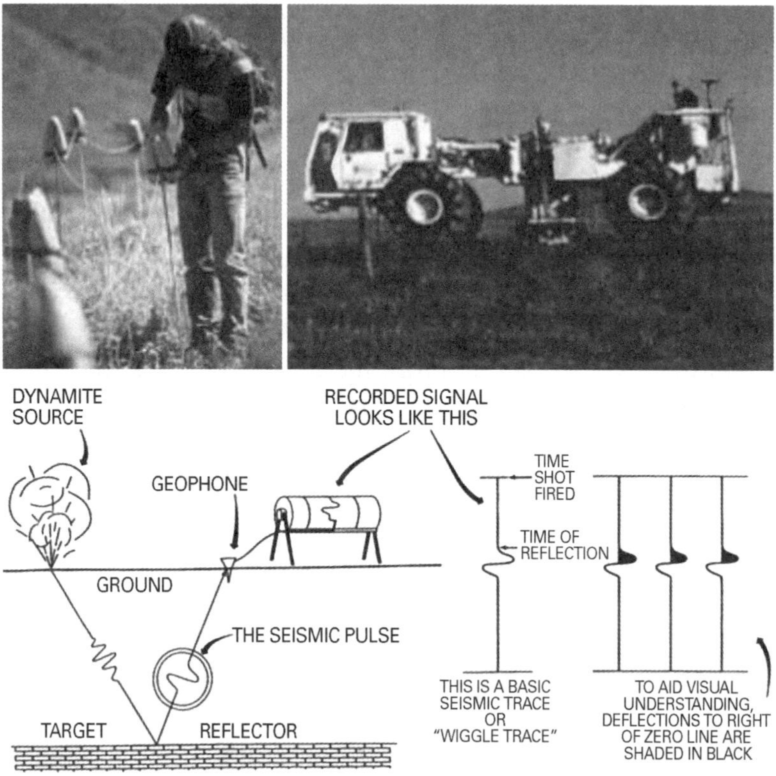

Fig. 7–2. Equipment and schematic in onshore exploration. *Source: API 2008, Natural Gas.org, 2008, Kessinger, 2008*

Technologies based on RFID and radio frequency can be used in gas exploration for managing both fixed and mobile assets. RFID-based markers can be used to supplement information available using GPS instruments. They can be used to mark and detect locations of current and past wells and geophone placement. Mobile asset management and tracking can be used to manage and track critical tools and parts issued to work crews at several locations.

Gas extraction. Gas extraction is the process of drilling exploratory wells to verify the presence of oil or gas, or both, and to determine the economic viability of reserves. By its very nature, it is a mobile equipment–intensive operation. The drill rigs are set up at locations determined in the field (fig. 7–3).

The equipment associated with drill rigs has to be managed and moved periodically (for onshore drilling). The supplies required during drilling have to be managed effectively. These include different types of drills, drill pipes, drilling fluids, fuel (for generators), etc. Additionally, the core samples collected during the drilling and the data on conditions encountered during drilling are analyzed continuously to help update the strategies to continue the drilling. The waste generated (such as waste drilling fluids removed from the mud, broken parts, and bits) have to be stored and shipped offsite.

Fig. 7–3. A mobile drilling rig. *Source: Anadarko Petroleum, 2007*

RFID-based technologies can and are being used to improve several aspects of gas exploration. These include management and automation of the daily or weekly shipment of supplies, tools, and parts to remote sites; tracking and management of the removal of wastes; and improving the process of returning equipment to storage yards and warehouses. The mobile asset tracking system discussed in chapter 3 can be used to process the orders received from the operator of each drill rig and aggregate the items into bins and pallets for periodic (such as weekly) shipment of parts to each site. By labeling tools, parts, and equipment, the warehouse is able to pick and aggregate them into bins/pallets, with each bin prepared for shipment to a site. Each bin is equipped with an RFID tag to help tag its shipment and track the receipt of the items in the bin.

Gas production at wells. The data collected from the exploratory wells is used to determine if the gas field and its wells are commercially viable. If a well is viable, it has to be "completed" for gas production. As part of the process of completion, the well is strengthened by installing casing. The casing is comprised of a series of metal tubes. Once the well intake structure is completed, a wellhead and monitoring equipment are installed to manage the gas flow to the production plant. The wellhead includes three components: casing head, tubing head, and "Christmas tree."

The *casing head* is comprised of heavy fittings to provide a seal between the well casing and the ground surface at the well. It is designed with sufficient strength to prevent a blowout by a buildup of pressures inside the well. The wellhead supports the entire length of casing in the well.

The *tubing head* provides a seal between the drill tubing that runs inside the casing and the pipe at surface. The gas or liquid flows out through this tubing.

The *Christmas tree* is the piece of equipment that fits on top of the casing and tubing heads (fig. 7–4). It includes tubes and valves to control the flow of hydrocarbons and other fluids out of the well. Its name is derived from its many branches that make it look like a tree. The branches have valves and monitoring equipment.

Fig. 7–4. Christmas tree equipment fitted to a wellhead. *Source: NGSA, 2008*

In a few instances, a natural gas well may need gas lifting equipment and treatment of the well to insure flow of gas. The treatment can include hydraulic fracturing of the rock layers and acid treatment of limestone beds to increase the flow of gas.

Oil may be present in the gas, especially if the gas is dissolved in oil under pressure in subsurface systems, or when gas is produced as a by-product from an oil well. When the pressure is relieved, the gas separates from the oil. This separation of gas and oil can be done by gravity in a closed tank known as a *conventional separator*. The oil is piped to a separate tank or system.

Gas heaters may be installed near the wellhead at locations where temperatures drop close to 0° Celsius. The water vapor can form crystals of ice and hydrates that accumulate at the valves and throttle the flow. The heaters prevent the formation of crystals of ice and hydrates. The heaters are operated by burning a small stream of the natural gas.

Sand present in the gas flow needs to be removed near the wellhead to prevent abrasion of valves, pipes, and downstream equipment. The sand is removed in sand scrubbers (filters).

RFID and wireless technologies can help improve the gas extraction operations at the well fields. The RFID chips can be used to label pipelines, parts, and valves to facilitate retrieval of information regarding the history of maintenance. RFID chips can also be used as primary and secondary markers installed around the gas field to locate abandoned exploration wells and fixtures, and to locate buried pipelines, including abandoned pipelines. The general application is discussed in chapter 5. An example of an application within the natural gas utility is discussed later in this chapter.

RFID and wireless technologies can also be a part of monitoring devices. These devices are equipped with data loggers and RF-based data transmission systems. The additional monitoring can help reduce the risks to operations (such as detecting changes in temperature and pressure in wells and on pipelines at 15-minute intervals). These monitoring devices also help reduce the life-cycle cost of equipment by improving the ability to track its maintenance. Along with the monitoring and management of the production from the well, RFID- and wireless-based monitoring devices help to increase the efficiency of wells.

Processing

The raw natural gas produced at the wells may contain by-products such as oils and heavier hydrocarbons. It may also contain gases such as hydrogen sulfide, carbon dioxide, and water vapor. These by-products and gases have to be removed before the gas achieves a quality that is acceptable for the transmission pipelines. The by-products can be collected and sold separately.

The raw natural gas is purified at natural gas processing plants. The plants incorporate four main processes, each of which will be discussed in more detail:

1. Oil and condensate removal
2. Water vapor removal
3. Extraction of natural gas liquids (NGLs) present in the natural gas
4. Hydrogen sulfide and carbon dioxide removal

Oil and condensate removal. The natural gas contained in the oil from oil wells may be separated in the well field using a conventional separator. After separation, the oil and gas are conveyed by separate pipelines to the processing plant.

In some instances, light crude oil and condensate have to be removed from the natural gas produced at natural gas wells. Low-temperature separators allow the gas to travel through a nozzle or choke mechanism which expands and cools the gas. Lowering the temperature of the gas removes the oil and water as condensates.

Water vapor removal. Water vapor is removed by passing the natural gas through adsorption towers containing dessicants such as silica gel or activated alumina. These towers are located downstream from compressors that pressurize the gas. The wet gas enters at the top of the tower and exits at the bottom. Two or more towers are located in parallel such that one can be offline to regenerate the wet dessicant while the other is still in operation. For regeneration, a heated air or natural gas stream is passed through the bed of dessicant. This vaporizes the water adsorbed to the surface of the dessicant.

Extraction of natural gas liquids (NGL). The natural gas, which is principally composed of methane, may contain heavier hydrocarbons such as propane, butane, pentane, and other heavier hydrocarbons. These are high-value hydrocarbons that need to be removed from the natural gas stream and recovered. To remove the hydrocarbons, the natural gas is passed through a bed of oil within an absorption tower. The oil becomes rich in hydrocarbons after passing through the gas and is then removed from the bottom of the tower. It is sent to a still where the hydrocarbons are boiled off by heating the oil to a temperature that is higher than the boiling point for the hydrocarbons but is less than the boiling point for the oil. The *lean oil* (with the hydrocarbons removed) is sent back to the absorption tower. The hydrocarbons are separated into their components (such as ethane, propane, and butane) by fractional distillation in a distillation tower.

Hydrogen sulfide and carbon dioxide removal. Natural gas is classified as *sour* (acidic) if it contains greater than 5.7 milligrams (mg) of hydrogen sulfide per cubic meter of gas. In addition to giving a rotten egg smell, the hydrogen sulfide generates sulfur dioxide when burned or oxidized in downstream processes, which ultimately forms sulfuric acid by combining with water vapor. This increases the corrosive nature of the gas. It can damage transmission and distribution systems. The hydrogen sulfide has to be removed prior to transmission.

The hydrogen sulfide is removed by conveying the sour gas through an absorption tower containing an amine solution (fig. 7–5). The amine solution extracts the hydrogen sulfide and carbon dioxide. The chemical process is called the *Girdler process.* The hydrogen sulfide can also be removed by passing the gas through a bed of iron filings/iron sponges. This removes the hydrogen sulfide as ferrous sulfide.

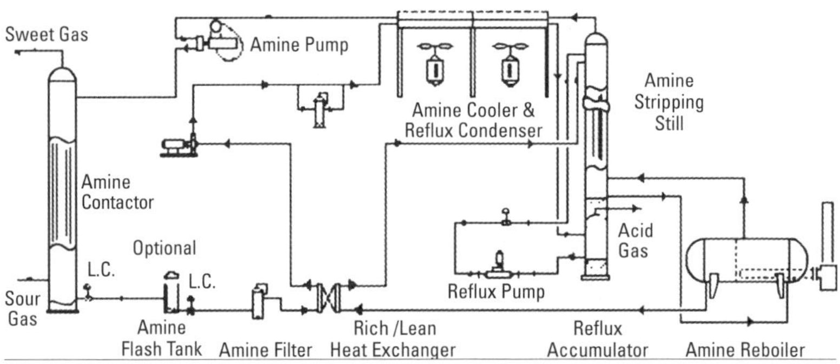

Fig. 7–5. Extraction of hydrogen sulfide and carbon dioxide using an amine solution. *Source: Newpoint Gas, 2007*

Further processing of the amine solution recovers the sulfur in an elemental form. The chemical process is called the *Claus process.* Figure 7–6 shows a sweetening plant (hydrogen sulfide and carbon dioxide removal). Figure 7–7 shows sulfur extracted as elemental sulfur at a gas production plant. The elemental sulfur is yellow in color.

Fig. 7–6. Natural gas sweetening plant (removal of hydrogen sulfide). *Source: Natural Gas.org, 2008; Duke Energy Gas Transmission, Canada, 2007; Spectra Energy, 2008*

Fig. 7–7. Production of sulfur at a natural gas processing plant (sulfur shown in the foreground). *Source: Natural Gas.org, 2008; Duke Energy Gas Transmission, Canada, 2007; Spectra Energy, 2008*

RFID and wireless applications at processing plants. RFID and wireless applications have a considerable number of applications within the gas processing plant. RFID chips can be used to label the locations of hidden fixed assets such as pipelines. The chips can be used to label and track the maintenance of critical fixed assets such as valves and switches. They can be used to label and track critical mobile assets such as tools, spare parts, and waste products. Automated metering technologies with radio frequency data transmission can be used to enhance the capabilities for continuous process monitoring. RFID-based access cards can be used to control and record access and improve the security at natural gas processing plants.

These applications are discussed in chapters 3 and 4. Specific applications in the natural gas industry are discussed in the following sections.

Transmission

Interstate pipelines. The processed natural gas is compressed and transmitted through high-pressure pipelines to storage reservoirs and local distribution companies (fig. 7–1). The gas is pressurized to a range between 200 and 1,500 pounds per square inch (psi) to reduce its volume for transmission. Additional compressor stations are located every 50 to 100 mi. (80 to 160 km) along the transmission pipeline to maintain the pressure and move the gas forward.

The transmission pipelines are typically 12 to 48 in. in diameter, although they may be any size. They are made of carbon steel and are coated with epoxy, polyethylene, or coal tar to prevent rusting and corrosion. At most locations, the pipelines are buried 36 to 48 in. below the ground.

Compression stations. The equipment in compression stations includes centrifugal or positive displacement (reciprocating) compressors. The motors that drive the compressors run off electricity or turbines powered by burning a portion of the natural gas. The compressor station may also include towers to dehydrate the gas (adsorption towers with dessicant) and the equipment to regenerate the dessicant (fig. 7–8). This equipment is similar to that in the processing plants. Additional scrubbers and filters may be installed to remove any hydrocarbons that condense out.

Fig. 7–8. Compression station along a transmission pipeline. *Source: National Grid, 2008*

Valves, metering, and control stations. Transmission pipelines are designed with major maintenance valves located every 5 to 10 mi. (8 to 16 km). These valves can be used to shut off the pipeline for periodic maintenance or emergency situations. Additional valves with motor operators may be installed along the pipeline to control and direct the flow of gas through the transmission network (fig. 7–9). The flow rate, temperature, pressure, and the operational status of valves are monitored and managed using a SCADA system in a gas control room that is manned and operated around the clock.

Fig. 7–9. Vertical natural gas filter and valve with remote operator. *Source: National Grid, 2008*

Storage

There are two types of storage that are part of natural gas systems (fig. 7–1):

1. Base load storage
2. Peak load storage

Base load storage. Base load storage is used to store gas during the low-demand summer season (April to October) and to boost supply during the high-demand winter season (November to March). Base storage helps the gas distribution system satisfy the variation in customer demand without having to increase the capacity of the infrastructure located upstream of the storage facilities (production and transmission pipelines located upstream of storage facilities).

In the United States, previously depleted gas reservoirs and aquifers are used for base load storage. The depleted gas reservoirs are gas fields that were part of producing fields in the past but whose yields have dropped below the point where they are economically viable. The advantage is that much of the infrastructure to inject and extract the gas is in place. In some instances, aquifers have been developed as underground reservoirs. However, unlike depleted gas reservoirs, aquifers can have leakage from the formation, and the gas leaking has to be collected by collector wells. Concerns with the risk of contaminating groundwater supplies and the higher losses associated with aquifers limit the use of aquifers for storage of natural gas.

It takes a gas utility a significant amount of time (several months) to fill and extract the gas from these base load storage facilities. Therefore, these facilities are used to meet the variation in demand on an annual cycle.

Peak load storage. Peak load storage is used to respond rapidly to changes in demand, usually brought on by changes in weather. The response time can be in hours for peak load storage facilities as compared to months for base load storage facilities.

The total volume available for peak load storage in the United States is much less than base load storage (less than 5% of base load storage). The principal method for peak load storage is in underground salt caverns. These caverns were formed by leaching out salt from within salt beds.

Aboveground storage. Natural gas can be liquefied and stored in aboveground tanks. Liquefied natural gas (LNG) occupies only 1/600th of the volume of natural gas. LNG can be generated by pressurizing the gas or cooling it at normal pressure to –260°F. LNG may be trucked to the storage site and transferred, or a liquefaction unit may compress natural gas from an adjacent pipeline into a liquid state and pump it into storage (fig. 7–10). The gas is vaporized back into the line or into a truck at times of higher demand (fig. 7–11).

Fig. 7–10. LNG storage tank. *Source: National Grid, 2008*

Fig. 7–11. LNG vaporizer. *Source: National Grid, 2008*

Additional storage for heavier hydrocarbons (NGLs). Additional storage tanks may be installed along the transmission system or in the distribution system downstream for storing heavier hydrocarbons in liquid form or as a pressurized gas. These are discussed below as components within the natural gas distribution and retail systems.

Application of RFID and wireless technologies

RFID chips can be used to identify the location of valves of underground pipes. The approach is discussed in chapter 5 and in the following section of this chapter. Primary markers with RFID chips can be used at aboveground locations. When the chip is scanned with a handheld device, its ID is correlated to information in a database stored in the handheld device. This is used to download the coordinates of secondary markers and the coordinates of points along the pipeline. The coordinates are stored both in GPS coordinates and relative to the location of the primary marker. The secondary markers can be used on stakes installed on the ground or on fixtures in the immediate vicinity of the pipeline. The frequency of coordinates (spacing between adjacent points along the pipeline) for which data is stored will depend on several factors. These factors include the level of accuracy desired, the curvature of bends, and the need to identify any deviations of the pipeline made during construction to avoid objects that were encountered.

Another application of RFID chips is for labeling valves. The chips are installed to provide an ID for the valve. The ID is read in the field by a handheld device. The device correlates the ID to a description of the type of valve, its function, and its maintenance history. Some critical information, such as the last date of maintenance, may also be stored in addition to the ID in the memory available in the chip.

RFID chips can be used to supplement the data transmitted through existing wireless networks. This technology enables additional data to be transmitted from remote locations. The information transmitted can include the condition of the asset, such as the temperature, flow rate, pressure, moisture, and constituents of the natural gas at different points along a line. The gathered and transmitted data can help keep track of the magnitude of losses along a line (through minor leaks) and help to identify the location of increase in leaks. It also helps with monitoring the condition of compressors and scrubbing and vapor removal devices.

Distribution

Pipelines. The natural gas is distributed to residential and small industrial customers through gas mains and service pipelines. The infrastructure for distribution is located downstream of the transportation (fig. 7–1). This infrastructure may be managed by the city, a local gas distribution company, a retailer company, or a major gas and electric utility.

The local distribution system uses a multitude of pipe diameters (from less than 1 in. to as large as 60 in.) in their pipeline systems. These lines are typically operated at pressures of 0.25 pounds per square inch gauge (psig) to 125 psig, much lower than the 600 psi or higher used in transmission systems. Traditionally, these pipelines were constructed of cast iron or bare steel. Today, the pipelines are made with polyethylene plastic or cathodically protected coated steel pipe (fig. 7–12). These materials offer a much longer reliable service life than cast iron or bare steel pipelines. Figure 7–13 shows an example where a utility has excavated a road and is inserting a plastic pipe inside a cast iron pipe.

Fig. 7–12. Installing gas distribution piping made of flexible plastic. *Source: National Grid, 2008*

Fig. 7–13. Insertion of plastic gas pipe inside an older cast iron gas pipe.
Source: National Grid, 2008

Distribution compressors and odor addition. In a few instances, small compressors may be used by the local distribution company (LDC) to maintain pressure in the distribution system (fig. 7–14). In some instances, the gas is scrubbed further to remove water vapor and condensates. Mercaptan is added to the natural gas to give it its odor (figs. 7–15 and 7–1). The presence of mercaptan facilitates leak detection within the distribution system.

Fig. 7–14. Distribution system compressor. *Source: National Grid, 2008*

Fig. 7–15. Odorant pumps. *Source: National Grid, 2008*

Metering. The service lines in the distribution system are connected to a gas meter and shutoff valve just before the gas line enters the home. Traditionally, the gas meter installed at the premises of each customer was read by visual observations (fig. 7–16). Today, gas utilities are incorporating various levels of automated meter reading using wireless technologies.

Fig. 7–16. Traditional gas meter (needs visual readout). *Source: National Grid, 2008*

NGL (propane, butane) distribution and retail. Propane customers either own or are provided with tanks at their site. The tanks are buried or are above ground and are piped to the dwelling or appliance (fig. 7–17). In some instances, a propane/air mixture is delivered through a network of underground pipes similar to a natural gas infrastructure.

Fig. 7–17. Installation of an underground propane gas storage tank. *Source: National Grid, 2008*

RFID and wireless applications—distribution and retail

There are significant opportunities for RFID and wireless applications in fixed asset location, identification, and tracking within the natural gas distribution and retail infrastructure. RFID-labeled primary and secondary markers can be used to improve the process of locating buried gas lines and other utility lines. The markers would store the coordinates of different points along the line as is discussed in chapter 5. The memory on the chip can also help store information on the nature of the line (metal, polyethylene plastic), the date of installation, and maintenance. Additionally, RFID-based labeling systems can help track the maintenance and replacement history of fixtures such as valves (such as dates of inspection, exercise, maintenance, and replacement of a valve).

RFID and wireless networks can increase the efficiency of monitoring and metering functions. Automated meter reading technologies use wireless transmission to wake up the chip on the meter by transmitting a signal from a handheld or vehicle-mounted device up to 30 m away. It may also transmit the data collected over a wireless network. This avoids the need to physically enter a property to read a meter.

Finally, RFID technologies are seeing applications in the tracking of tools, equipment, and vehicles that are part of a utility in charge of maintaining the gas distribution system. The large number of vehicles, together with the geographic layout of the distribution system that may encompass several cities, makes it advantageous to implement a mobile asset tracking system. Mobile asset tracking is discussed in chapter 4 and in the following section.

Mobile assets

Tools and equipment. Each company responsible for the distribution system stores mobile assets required for maintenance at one or more warehouses. These assets include the following:

1. *High-value tools for maintenance and troubleshooting.* Examples are specialized monitoring equipment such as flow meters, leak detectors (fig. 7–18), laptop computers with interfaces to hook up such sensors, pavement breakers, etc.
2. *Small gear and supplies.* Items such as gloves, helmets, boots, etc.
3. *Parts and spares for larger equipment.* Items such as compressors, valves, and meters.

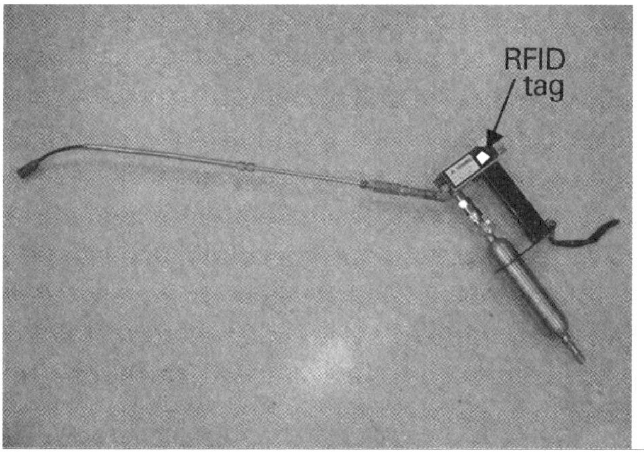

Fig. 7–18. A gas leak detector. The RFID tag can be installed adjacent to the bar code. *Source: National Grid, 2008*

When a work order is issued to a service (maintenance) crew to conduct a repair, the crew members may visit a warehouse or storage area to obtain the material required to conduct the repair. Each work order may require items from different stock keeping units (SKUs). An example is a work order for a compressor station. The crew may need gloves, mineral oil for removing rust, a specialty wrench that is to be issued from a tool crib and returned, and process-monitoring equipment such as a flow meter and handheld computer that are to be returned after the work order is complete. The items required are collected off the shelves by the staff at the warehouse and brought to the checkout counter where they are issued to the maintenance crew. In some instances, the maintenance person may walk through a section of the warehouse and pick out the specialty equipment required for the task.

RFID technologies can be applied together with bar code scanners to improve the management of such equipment needed for work orders. The goals of the RFID-based system are the following:

- Improved utilization of specialty equipment
- Improved management of inventory to prevent delays (from lack of availability of specialty equipment/stockouts)
- Minimization of the time that the equipment is sitting in inventory
- Reduced shrinkage (loss due to theft or misplacement)

Fleet. Gas transmission and distribution utilities maintain a fleet of vehicles that the work crews use to inspect and service the equipment at various facilities (such as compression stations, tanks, valves, and SCADA telemetry). The fleet includes small to large trucks, earth moving equipment of different sizes, and cranes and forklifts.

These vehicles are parked at maintenance or vehicle yards. Once a maintenance work order enters the system, the staff assigned to the work order may request specialized vehicles required to complete the work order. The vehicle fleet manager has to make these vehicles available with minimal delay.

The vehicle fleet manager has to track the usage of the vehicles. He has to balance the competing demands of reducing the number of vehicles that are idle (which requires reducing the size of the fleet) and minimizing the delays due to unavailability of vehicles. (The latter requires increasing the size of the fleet or insuring access to vehicles that can be rented from other parties). In addition to tracking vehicle usage and delays due to unavailability, the manager has to track the maintenance schedule and the consumption of parts and fuel, as well as make decisions regarding types of vehicles and parts to order in the future.

The application of RFID-based technology and enterprise software allows the fleet manager to maximize the time a vehicle is utilized. They can help improve the workflow and improve the utilization and operation of vehicles and maintenance yards. The general applications are discussed earlier in chapters 1 through 5; specific applications within the natural gas industry are discussed in this chapter.

Monitoring

The companies involved in various parts of the natural gas industry have to monitor and maintain assets that are geographically distributed. Many of these facilities are managed remotely (fig. 7–9). The utility may have a control system such as a SCADA system that allows several facilities to be monitored and operated remotely.

Within a SCADA system, the readings from sensors monitoring the status of equipment are converted to electronic signals and transmitted to a central location. The method of signal transmission can be hardwired or wireless. The person or persons monitoring the SCADA system are notified of abnormal operating conditions through alarms. Depending on the nature of the alarm, the employee monitoring the SCADA system calls in personnel to service the alarm. The SCADA system is also used to troubleshoot and remediate problems (such as turning off valves, raising or lowering pressure or flow rates, or a compressor running at a temperature higher than is normal) from a remote location.

The presence of an IT backbone servicing a SCADA system makes it easier to incorporate RFID- and GPS-based technologies. An example is use of a chip that can store 120 days' worth of hourly flow data from a water meter and transmit the data when activated by a signal from a handheld device that is 30 m (100 ft) away. This allows the utility to not only obtain a single water meter reading but also provides information on the pattern of usage by customers, aids in leak detection, and helps to prioritize repairs.

Analysis of Problems and Solutions

This section describes five categories of problems for which solutions can be developed using RFID and wireless technologies. These five categories are presented as a matrix in table 7–1. The matrix helps relate processes in the natural gas industry to problems faced.

Table 7–1. Categories of problems that can be addressed with RFID and wireless technologies at gas utilities

Gas Industry Process #	Description of Assets	Categories of Problems				
		Fixed Assets	Mobile Assets	Vehicle Tracking, Fleet Management	Secure Access	Monitoring/Recording
1. Extraction	Gas fields, exploration wells, extraction wells, scrubbers and separators in the field	Location + status of wells, location of underground pipelines	Equipment checked out for maintenance, loaded in service vehicles, rigs	Seismic trucks, Service vehicles tracking, utilization, parts & maintenance		Well pressure, temperature of lines, air temperature, gas constituents
2. Production Plant	Compressors, adsorption towers and absorption towers for condensate, vapor, hydrogen sulfide, CO_2 removal, separation of natural gas liquids	Location of underground and hidden pipelines, legacy pipelines from earlier processes	Monitoring equipment, maintenance equipment	Service vehicles	Certain equipment, fuel and storage areas, gas production areas	Maintenance and replacement records on equipment; data logging and wireless transmission of gas constituents, emissions, temperature, vibration, moisture, etc.
3. Gas Transmission	Underground lines, compressor stations, data networks, storage areas	Location, history of maintenance,	Same as 2	Same as 1	Compressor stations, storage areas	Same as 2
4. Gas Distribution and Retail	Underground gas mains and service lines, storage areas, gas processing plants (to add odors), compression and scrubbing stations	Location of assets, maintenance history, assets of other utilities and their pipelines	Same as 2	Same as 1, excavators	Gas storage, processing areas	Same as 3 + leak detection
5. Centralized Operations	Warehouses, vehicle fleets		Same as 2 + Inventory	Same as 1		

Fixed assets

Description of problems—locating assets. Natural gas utilities continue to face problems with locating aging fixed assets. These include old and existing gas mains and service lines that are located underground in the vicinity of buildings.

Two recurring problems are the unearthing of old lines and the potential for damage to adjacent lines during the excavation to access and repair a pipeline. This is due to the proximity of lines to other structures and utility lines, their colocation within the same vertical plane, and their deviations from the locations shown in the drawings.

The challenge in developing a solution to this problem is the high level of accuracy required for locating the lines. Also, the solution must involve a method of incorporating information on the location of lines from the drawings into a database that has a graphical interface, together with an ability to update the information gained from additional field surveys and explorations. There has to be a method for incorporating discoveries made during an exploratory survey into a database for locating assets. The information from this database has to be accessible in the field.

Solution for locating assets. This section discusses how a solution can be developed using an RFID-based permanent staking and marking system. One solution under consideration is the installation of markers (labels) embedded with RFID chips at the corners of buildings, on fixtures, or at locations along the edge of a tank, etc. The coordinates of points along a line are stored relative to the locations of the primary and secondary RFID labels. The primary labels are used to locate the secondary labels. The secondary labels are located on or in the vicinity of assets. This information is accessed through a handheld device. The coordinates locate various points along each pipeline (chapter 5). The spacing between adjacent points along a pipeline stored in the database is based on several parameters. These parameters include the following:

- Location of adjacent ends of each straight section of pipeline.

- Locations where adjacent lines cross over each other or come close to each other.
- The number of points required to describe the curvature of the pipeline at its bends. (The number of points selected should fully describe the arc.)

The chips selected as part of the labeling system should survive over an extended period of time (20 years or longer). The locations of the markers should be such that they are permanent (not damaged or moved during maintenance or construction). Only when major construction is done would the markers and labels be removed and replaced by additional locations that are labeled. The labeling system must have a level of redundancy such that when a point is lost (due to storm damage, construction damage, a chip goes out of service, etc.) the location of the pipeline can be determined by the coordinates derived from other markers.

Description of problem with identification and status of assets. A second problem facing field crews is the identification of similar assets. For example, when a crew is in the field, it takes considerable effort to distinguish one valve along a pipeline from another in the vicinity unless metal tags are installed and stored in the roadway box. If there are parallel underground pipelines, the crew may have difficulty distinguishing the lines and the valves that correspond to the lines. For example, there may be a possibility that lines may cross each other underground at some distance from where the valves are located.

Even after the assets have been located (e.g., valves on parallel pipes), the crew is unable to determine or verify when the last repair was conducted unless the information was entered into a database. In some instances, the record may differ from the condition observed in the field (the valve appears newer or older than what is in the database).

Solution for identification and status of assets. This problem can be overcome if the information on the repair can be stored by the crew in an RFID chip placed on the asset or in the vicinity of the asset. The memory on the chip can be used to store the last repair or service date. A level of redundancy can be maintained by also storing this information on a database. The data is entered in a handheld device in the field and is synchronized with the enterprise software system at the office at the end of the day. The operator can log into the database using an ID or an RFID-based access card. When the vault or structure housing the valves is accessed in the future, the crew can obtain a confirmation of the maintenance record (such as the last maintenance date) by using a handheld device to read the information stored in the memory of the RFID chip.

Figure 7–19 shows how a labeling system can be implemented at an LNG vaporizer site. Figure 7–20 shows the same system in a building. In these instances, the secondary labels are not for locating the asset. They are installed to read and verify the maintenance date on the asset. This helps the maintenance or inspection crew determine when the asset was last maintained.

Fig. 7–19. Primary and secondary label installation (RFID tags) on an LNG vaporizer plant

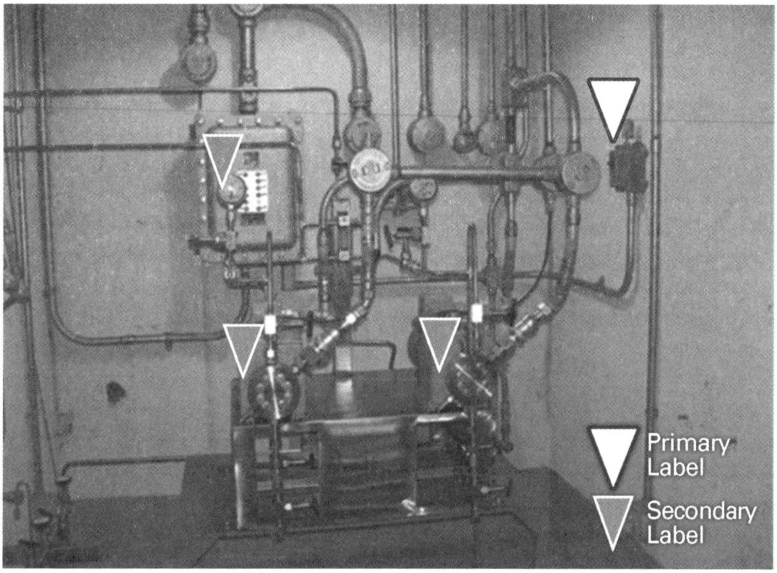

Fig. 7–20. Primary and secondary label installation (RFID tags) inside a building

The primary label is located at an easily accessible point, such as near a light switch. Scanning the primary label will inform the technician about the number and the locations of the secondary label in the building or in the vicinity of the primary label. The function of the primary label is similar to a map showing a list of secondary labels. Having access to the primary label or a map is important if a person is doing an inspection. The primary label (or the map) is necessary if the person is new and is not familiar with the location of all the secondary labels in the building. Without information on the locations of all the secondary labels, the employee could miss certain secondary labels, and the inspection would be incomplete.

Another possible scenario occurs when an outside electrician (or technician) needs to perform an inspection and possibly a repair. While the electrician could carry a map or a list of the labels and their locations, in some instances, this may not be readily available. The primary label would help the electrician determine if the asset of interest is labeled and then go to the asset, rather than the electrician returning to the main office to print a map. The electrician can go to the secondary label or labels of interest and scan the information. This will provide information regarding the last date of maintenance. If the electrician completes a maintenance work order on the asset, he or she updates the information stored in the memory of the tag, in addition to updating the database in the office.

Mobile assets

Description of problems with tool tracking. Each gas utility has a certain number of mobile assets that are critical for troubleshooting problems. Examples can include handheld computers, vibration sensors, temperature sensors, data loggers, and total stations; also tools that are part of a tool crib, such as specialty wrenches and drills. These units are housed at a few locations (such as warehouses) within the service area of the water utility.

Once a service crew is assigned a work order to repair a piece of equipment (length of pipe, valves, compressors, scrubbers, etc.), its members identify what tools are necessary to conduct the repair. They go to the warehouse and request the equipment. If the warehouse staff is unable to track critical tools, it results in delays to work orders.

To make this process efficient, the warehouse staff should be capable of making following decisions:

1. Immediately determine whether all the tools necessary for the repair are available.
2. If some tools are unavailable, determine the time at which they will be returned to the warehouse and become available.
3. Determine the condition of the tools, both at the time when they are being returned to the warehouse and when they are reissued.
4. Determine if the tools need to be repaired before they are reissued.
5. Set aside and send damaged tools for repairs or order replacement tools.

6. Determine the number of tools of each type that need to be in stock to prevent delays for the work crews (size of inventory required).
7. Determine the best location to store tools (which warehouse should store which tools and how many).
8. Reduce the amount of time tools are sitting unutilized (both with the service crews and on the shelf).
9. Evaluate techniques to reduce loss and misplacement of tools.

Developing RFID-based solutions for tool tracking. Implementing mobile asset management within the gas utility is simpler than applications at high-volume industries such as retail. For example:

1. The number of critical assets that need to be tracked is limited (typically, less than 100). These consist of moderate- to high-value mobile equipment.
2. The number of units of each critical asset is limited. There may be 1 to 500 units in circulation within a water utility (e.g., a handheld computer).
3. Units are housed at a few locations.

In implementing a solution, the first step is to identify those mobile assets that are critical to the operation of the utility. For example, a decision can be made to label only those units that are of high value:

1. A handheld scanner can be used to read tags at the time the high-value items are issued and returned. This can be supplemented with an additional bar coding or detection system.

2. Scanning can be done on a per unit basis because there is limited throughput—only a few units are issued each day. This allows opportunities to minimize problems with interference that occur in high-throughput systems when multiple units (of different types) are passing in front of the scanner within a very short time interval.

This strategy has been used as part of solutions for mobile asset tracking that are in the market (e.g., Houndware by Verichip). The solutions use a combination of bar codes and RFID chips. Once the asset tracking system is in place, the software system is set up to improve the management of the assets.

Problems with fleet management. A medium to large gas utility may maintain a fleet of 100 to 1,000 vehicles. These include:

1. Small vehicles—cars, small trucks
2. Four-wheel-drive vehicles used at job sites
3. Heavy equipment and large trucks—trucks, excavators, cranes, water trucks, etc.

The fleet may be located at the various maintenance yards distributed around the service region. In some instances, the service area of gas utilities spans multiple states. Some vehicles are assigned to work crews. Others are assigned on the basis of assignment of work orders. The challenges facing gas utilities include the following:

1. Effective utilization of vehicles (reducing the idle time).
2. Management of spare parts to improve utilization and reduce fleet maintenance costs. This includes tracking the cost and longevity of parts, optimizing their inventory, and ordering parts in a timely manner.

3. Tracking fuel consumption.
4. Scheduling service.

Developing solutions for tracking vehicles as part of the maintenance fleet. RFID-based labeling presents unique opportunities that extend the capabilities for fleet management beyond what is possible with bar codes. One can label parts or vehicles at locations (within the vehicle) that are not able to be scanned with a bar code. For example, the chip may be located on the inside of the windshield of a car where it can be read by a scanner installed at the entry or exit points of a vehicle yard. An example is the RFID-based FastPass system that allows a car to pass through a toll booth with automatic payment. The utility may use a similar system to log when a vehicle leaves and reenters a parking lot, maintenance yard, or a plant. Another application of RFID technology is the labeling of an engine or a part in the car, such as a tire hub, to log the history of maintenance and replacement of parts. It may also be a part of car keys and card readers assigned to keys (as in Speedpass by ExxonMobil, Smart keys by Nissan and other car manufacturers).

The vehicle fleet management staff can use the RFID-based system and software to accomplish a wide variety of tracking tasks:

1. Track a vehicle—correlate the operator or maintenance crew to a vehicle
2. Track where the vehicle is parked
3. Track maintenance and parts used on each vehicle
4. Track fuel usage by vehicle
5. Track the life cycle of parts
6. Track the effectiveness of preventative maintenance

The fleet manager can then apply the information to better manage the frequency for ordering parts. If there are multiple suppliers for parts of the fleet, the tracking system may be utilized to determine which suppliers' products are more suitable for the purposes of the utility.

Chapter 4 discusses additional applications of RFID and wireless technologies as part of GPS- and satellite-based vehicle tracking systems. These can be used by a water utility that services a large geographic area.

Problems with material tracking. Gas extraction fields and production plants generate a variety of material that could benefit from a tracking/labeling system similar to that used for tracking hazardous wastes. For example, a gas production plant may be storing quantities of chemicals for removal of hydrogen sulfide. An extraction facility may be storing used drilling fluids. These may be stored in containers or drums for disposal at a later time. Without a tracking system, it is difficult to determine the type of waste in a container or a storage area when a hauler arrives at the site several months (or years) later.

Another example is the tracking of small quantities of waste material, some of which may be hazardous. The quantities may be generated in small quantities but can accumulate over time. Without a continuous tracking system, it is a challenge to identify the date or dates on which the material was generated and the treatment processes that generated the material, and to determine whether it is hazardous or not. Subsequent tests to identify whether the material is hazardous are considerably more expensive.

Solutions for material tracking. The use of RFID chips to label drums and containers, however small, can provide an attractive solution for materials tracking and inventory management. Each bottle or drum can be labeled with a chip that has a unique ID number. The operator may program in a code that describes the waste into the memory of the chip and store it in a database. When the waste has to be identified prior to disposal, the operator has to scan the chip to obtain the requisite information regarding the source and nature of the waste.

Access management and infrastructure security. Most gas utilities have key-based access to their facilities (i.e., they use keys rather than access cards to open doors to buildings). This is a reliable means of access that has proven itself over the years. The limitations with key-based access are as follows:

1. There is no log describing who accessed a restricted facility at what time.
2. For facilities that need routine inspection, a separate log must be created for the inspection.
3. There is no log of the vehicles that entered facility and the types of wastes brought in or carried out of a facility.

Installing a card reader can help address several of these limitations such as the following:

1. Maintaining a log of access (important for passing inspection audits)
2. Ability to integrate information from the reading on an instrument to the operator
3. Controlling the access, if required

RFID-based solutions can be implemented as an alternative to card readers. For example, a utility may implement the following to maintain a record of wastes hauled into or out of a facility:

1. Install RFID chips on all vehicles that may access a facility (on the fleet used by the utility and the haulers' fleet).
2. Automatically record which vehicle accessed the facility.
3. Require the vehicle operator to scan a badge or enter a number to open a valve, switch

adsorption and scrubbing towers that are in operation, or to haul waste out of a facility.

Monitoring and recording

Challenges in monitoring and recording. Gas production, transmission, and distribution companies are seeking to improve the monitoring and recording of processes to reduce operating costs, improve maintenance procedures, and satisfy regulatory requirements. For example, gas extraction engineers prefer to have more frequent data on wells to identify strategies to continue the drilling. Gas production and transmission companies may want more frequent data on moisture, hydrogen sulfide, and pressure along the line to identify if scrubbers and compressors are operating properly. While the cost of monitoring continues to be a challenge, recent advances in RFID can help enhance the capabilities to transmit data from monitoring devices into a SCADA system.

Solutions for monitoring. The integration of RFID chips with other metering and data recording systems presents some unique opportunities to increase monitoring and lower costs. RFID chips can be integrated with measuring devices (such as temperature sensor, pressure sensor, gas flow meter, or gas constituent detector) and data loggers. The latter are capable of storing several measurements in their memory or on the chips. The RFID chips can transmit data using a wireless system and at more frequent intervals.

For gas distribution companies, the RFID chips can be part of automatic meter readers and data loggers. The chip can be activated remotely and instructed to transmit data. This capability is being used by gas utilities to obtain periodic records (monthly to hourly) of gas consumption. It can also be used to identify the magnitude of losses along lines, track the pressure, and identify loss of pressure.

Solutions for data recording. RFID chips can be used to label assets in a manner that improves the management of the asset. The memory of the chip may contain data that serves as a unique identifier. In this instance, a handheld computer has to use this identifier to access a database that contains a history of maintenance on the asset. The database may reside on an enterprise system. The latest data from the database may be downloaded to the handheld device. The same functionality can be incorporated using a bar code except for the fact that the RFID chip does not require direct line of sight. The asset would still be tagged with a unique ID in case the chip fails.

In a more advanced application, the memory chip could be used to program the last service date, in addition to its unique identifier. This frees up the need to access a database.

The types of data records that may benefit from such a system include the following:

1. Finding a record of data from earlier surveys (e.g., abandoned wells or drills)
2. A record of date and type of activity for maintenance performed on an asset
    - Repair of equipment
    - Oil change

The system may be designed to provide better access to the information in the field to help with decision making.

Costs and Benefits and Improving the Return on Invested Capital

The most desirable outcome of an RFID and wireless technology implementation is a low-cost, high-benefit, large return on the invested capital. For this, the authors recommend that the utility conduct an asset management plan through which it identifies the most critical problems. These are problems that disrupt the normal functioning of the utility, pose significant risk to personnel, or result in a substantial cost or liability. Once the problems are prioritized, the next step is to move to implementation.

The implementation of RFID- and wireless-based solutions requires an understanding of the hardware and software that are discussed in chapters 1 to 4. It is recommended that the solutions to the categories of problems discussed be implemented as separate or as phased/sequential projects. Certain projects such as those for monitoring can be implemented as part of plant-wide instrumentation projects. Others such as mobile asset management can be implemented within a warehouse. The vehicle management can be initiated at vehicle maintenance, storage, and refueling areas. The static asset management can be implemented as part of a plant expansion.

Appendix A discusses costs and benefits that go into an ROI computation around an RFID-based solution for asset management.

References

1. Anadarko Petroleum. 2007.
2. API, American Petroleum Institute. 2008.
3. Duke Energy Gas Transmission, Canada, 2007.
4. Kessinger, W. 2008. www.walterkessinger.com. Seismic data acquisition.
5. National Grid. 2008. (Figures 7–8 to 7–18.)
6. Natural Gas.org. 2008. From wellhead to burner tip.
7. NGSA (Natural Gas Supply Association). 2008.
8. Newpoint Gas. 2007. Gas Treating Services.
9. Spectra Energy, 2008.

A EVALUATING THE BUSINESS PROPOSITION OF AN RFID PROJECT

Evaluation of an RFID Project

Before embarking on an RFID project, it is important to think through many factors to determine the feasibility of the project. It is always advisable to start with a small pilot project, learn from it, and then scale up with a bigger implementation.

One should start by considering the nature and purpose of the application. Is it for compliance with a customer such as the Department of Defense or a government regulation? Is it for business process efficiency? If it is an application that can result in benefits such as cost savings or increased revenue, one may be able to come up with a clear model for demonstrating *return on investment (ROI)* [also termed *return on invested capital (ROIC)*].

One should perform extensive research to determine what type of RFID technology would be appropriate, based on the desired read range, environmental factors, data volume, and cost. RFID software and hardware vendors should be able to provide white papers, presentations, and demos explaining the cost, benefits, and possible applications of the technology they sell and how other customers have evaluated and used their solutions. Sometimes, funding for pilot projects may be available from government agencies or trade organizations.

The following is a list of points to consider:

- **Determine whether to use active tags or passive tags.** The following must be taken into account:
    - *The required read range.*
    - *Environmental conditions.* These include the temperature of the surroundings, the presence of metals, and whether it is necessary to read in a liquid medium.
    - *Number of tags.* This relates to the number of tags that would be needed per week, month, etc.
    - *Reuse of tags.* Whether or not the tags can be reused must be evaluated.
    - *The desired longevity of the tags.* Active tags are powered by batteries that run out after a few years, while passive tags can last forever as long as they are not damaged.
    - *The type of packaging and form factor of the tags needed.*
- **Determine the type of readers to use, stationary or portable.** Handheld readers are good if the volume of tags is not very high, since a person is physically required to operate them. Stationary readers are recommended if one needs to cover a certain area automatically such that any tagged item within its range is detected. One should consider the number of tags that can be within a reader's range at a given time and also the area that needs to be covered.
- **Determine if RFID middleware software is required, and if so, its desired functionality.** The following points should be considered:
    - *One vendor versus multiple vendors.* Is it expected to have readers from different vendors or one vendor? If there will be

multiple vendors, it is easier to have a middleware that has the capability of communicating with readers having different APIs.

- *Volume of data.* The volume of data that is generated by the RFID readers and the kind of event filtering required must be considered in order to avoid overwhelming the receiving applications.

- **Evaluate the current IT infrastructure.** It will be necessary to determine to what extent existing hardware can be utilized for the pilot project and to determine if the existing network will be capable of handling the increased data traffic. If not, what will be the additional networking cost?
- **Consider the data use.** How will the data generated by the RFID system be used? Which applications will be receiving the RFID events? Are they capable of utilizing the information? One of the main uses of an RFID system is to allow item-level information, which all existing applications may not be able to handle in their current state.
- **Keep in mind that an RFID system can generate a lot of data.** There has to be a strategy in place as to how long the data should be kept in the database and what happens to it after that.
- **Consider if the data or events will be shared and decide how this will be carried out.** Will it be necessary to share the output data or events from the RFID system with applications that reside with other companies (vendors or partners)? How will the existing processes for sharing data with such companies change as a result of the new RFID system?

- **Consider the RFID system's compatibility with existing processes.** How will the RFID system coexist with processes that are already in place today? It is not advisable to prematurely change the existing processes until the pilot is successfully completed, the ROI analysis is performed, and the value of scaling the solution is clear.
- **Evaluate documentation and training.** Along with the implementation, it may be necessary to have new documentation and also to train the staff so they are able to use the new system. Determine the complexity and time involved in doing that and how it may impact the operations during the transition.
- **Consider the security and privacy aspects of the application.** The security and privacy aspects of the RFID system also need to be carefully evaluated. Since data can be stored at different locations in an RFID system, including the RFID tags, RFID readers, the RFID middleware, and receiving applications, care must be taken to prevent unauthorized access and other security threats. If the RFID system involves the consumer, they should clearly understand the implications of such use.

For larger projects, one should prepare a project plan outlining the cost and the benefits of the pilot project. Benefits can include improved customer service, supply chain efficiencies, inventory optimization, improved safety, fewer accidents and errors, reduced warehouse cost, reduced theft, reduced time for operations, and more.

Based on the cost, benefits, and time required, create an ROI model. After the pilot is completed, determine if the ROI is sufficient to justify scaling the project and accordingly plan the next steps.

Understanding Risks Associated with an RFID System

For RFID implementations to be successful, organizations should effectively manage their risk. The major categories of risk are as follows:

- **Business process risk.** This encompasses threats and vulnerabilities that could cause part or all of the RFID system to fail. Potential impacts range from a deceleration of the business process to the loss of critical business or operational records. Business process risk can occur for many reasons, including human action (either benign or malicious) and natural causes. Factors influencing business process risk include the importance of the RFID-supported business processes to the mission of the organization, the robustness of business continuity planning, and the environment in which the RFID technology is located. The cloning of tags and attacks on enterprise subsystem networks are examples of threats to business processes.
- **Read rate for RFID tags.** Factors that affect the radio frequency used include the following:
    - Industrial usage of frequency spectrum in the range of the frequency being used for RF communications
    - Presence of liquids, chemicals, and metals (absorption of radio waves and reflection of radio waves)
    - RF communication distance, speed, and directivity
    - Anticollision and transmission protocols
- **Privacy risk.** Privacy rights or expectations may be compromised if an RFID system uses what is

considered personal information for a purpose other than originally intended or if a third party uses the presence of tagged items to profile individuals. In the case of the latter, the primary privacy risk is likely borne by the consumer, not the organization that implemented the RFID system. Nevertheless, the RFID implementing organization still has privacy-related risks, including penalties from noncompliance with existing privacy regulations, legal liability, and the reaction of consumers, employees, public interest groups, and other stakeholders.

- **Externality risk.** Every connection point between an RFID system and other systems represents a potential vulnerability. One externality risk for an RF subsystem is due to the hazards resulting from electromagnetic radiation, which could possible range from adverse human health effects to ignition of combustible material, such as fuel or ordnance. The main externality risk for an enterprise subsystem is successful attacks on networked hosts and applications. Computer network attacks can involve *malware,* or attack tools that exploit software vulnerabilities and configuration weaknesses to gain access to systems, perform a denial of service, or cause other damage. The impact of computer network attacks can range from performance degradation to complete compromise of a mission-critical application.

Improving the Return on Invested Capital (ROIC)

The most desirable outcome of an RFID and wireless technology implementation is a low-cost and high-benefit project with a substantial return on the invested capital (ROIC). To achieve this end, the authors recommend that the utility conduct an asset management plan through which it identifies the most critical problems. For this, it is necessary to first identify the problems around the following:

- Tracking the history of maintenance on assets
- Tracking the mobile equipment and vehicles
- Addressing problems with discerning the location of underground or hidden assets and the history of potential for damage to such assets
- Need for higher levels of infrastructure access control and infrastructure security
- Addressing difficulties and bottlenecks around the implementation of continuous usage monitoring (water, electricity, gas usage)
- Monitoring the status of remote equipment (such as temperature, current drawn, air flow, etc).

Once the problems are identified, one may prioritize them based on loss of revenue, liability (including health and safety), regulatory compliance, and opportunities to reduce costs. This helps identify the most critical problems. The next step is to move towards developing a scope of work for implementation.

The scope of work for implementing an asset management program that encompasses RFID- and wireless-based solutions requires an understanding of the hardware and software discussed in chapters 1 to 4. It is recommended that the solutions to the five categories of problems discussed in the remaining chapters then be evaluated as separate or as phased/ sequential projects and as opportunistic projects that leverage planned investment in IT infrastructure. For example, certain projects such as those for monitoring can be implemented as part of plantwide instrumentation projects, thereby leveraging the investment on such projects. Others such as mobile asset management can be implemented where the need is the highest, such as within a warehouse. The vehicle management can be initiated at vehicle maintenance, storage, and refueling areas. The static asset management can be implemented as part of a plant expansion.

Table A–1 contains a list of costs and benefits that go into the computation of a return on invested capital (ROIC) computation. This table is provided as a guide for managers considering the implementation.

Table A–1. Costs, benefits, and ROIC computation for RFID implementation

Determine the costs and benefits for implementing RFID and wireless technology based solution for each of the five categories:

1. Fixed (hidden and visible) asset management;
2. Mobile asset management (tools, spare parts);
3. Vehicle and fleet management;
4. Secure access; and
5. Monitoring (automated meter reading, continuous equipment monitoring).

Document Costs:
  Hardware:
    Chips, scanners, computers, wireless transmission antennae

  Software
    Middleware, enterprise software

Document Benefits:
  Higher knowledge of maintenance history
  Lower life cycle cost on equipment
  Reduced risk
  Higher utilization
  Lower rate of loss
  Better management of inventory
  Higher customer service

Compute the return on invested capital (ROIC)

One of the challenges in computing the ROIC is the determination of the benefits of reduced risk. Some approaches that may help are historical estimates of risk at the utility or in similar industries. The risk manager may have a historical estimate of risks. The risk is related to preventable accidents and the cost incurred from such accidents. Typically, the risk increases with time, with the age of the facilities, and with the addition of infrastructure within a certain geographic area. The manager may use his or her judgment to compute the cost of such accidents. This may have to be estimated from incidents that occurred at other locations around the world.

Where it becomes difficult to estimate ROIC, one may approach the problem from other benefits, opportunities, or regulatory drivers. The opportunities may be to add RFID-based solutions to planned IT infrastructure projects. The principal costs may be those associated with the implementation of a SCADA system. The RFID-based extension is an additional component to the project that is a small part of the overall cost.

Table A–2 is a recommended series of steps for implementation. The table should help increase the benefits from an RFID implementation.

Table A–2. Recommended steps for implementation

1. Determine the highest pain points (highest risk or maintenance problems) and the most critical problems
2. Determine the location within the infrastructure with the highest pain points
3. Determine the location where existing IT infrastructure is to be upgraded
4. Implement a pilot
   a. The pilot may be at a location where it is easier to implement and verify the benefits
   b. It may not be at a location with the highest pain point
   c. It may be at a location where IT infrastructure is available or is being upgraded
5. Work out all the bugs
6. Improve the algorithms
7. Expand to the location with the highest pain points
8. Expand to the facility or the Utility

B Industry Acronyms and Abbreviations

ACL	asynchronous connectionless
AI	antenna interface
AIDC	automated identification and data capture
AIT	automated information technologies
AMPS	advanced mobile phone system
ANSI	American National Standards Institute
AP	access point
API	application interface
API	American Petroleum Institute
ASCII	American Standard Code for Information Interchange
ASN-GW	access service network gateway
BS	base station
BSS	basic service set
BT	Bluetooth
BWA	broadband wireless access
CDMA	code division multiple access
CDPD	cellular digital packet data
CEP	complex event processing
CSMA/CA	carrier sense multiple access with collision avoidance (protocol)
CSMA/CD	carrier sense multiple access with collision detection (protocol)
CSN	core service network
CSV	comma separated variable
DNS	Domain Name Service
DOD	Department of Defense

DSSS	direct sequence spread spectrum
DST	Digital Signature Transponder
ECMA	European Computer Manufacturers Association
EDGE	Enhanced Data Rates for GSM Evolution (standard)
EDI	Electronic Data Interchange
EDR	enhanced data rates
EPC	Electronic Product Code
EPCIS	EPCglobal Network EPC Information Service Specification
ERM	enterprise resource management
ERP	enterprise resource planning
ETSI	European Telecommunications Standards Institute
FA	foreign agent
FCC	Federal Communications Commission
FEFO	first expired first out
FHSS	frequency hopping spread spectrum
FIFO	first in first out
GB	gigabytes
GGSN	gateway GPRS support node
GHz	gigahertz
GMSC	gateway mobile switching center
GPR	ground penetrating radar
GPRS	general packet radio service
GPS	global positioning system
GSM	Global System for Mobile Communications (standard)
HA	home agent
hazmat	hazardous materials
HF	high frequency
HLR	home location register
HSDPA	High Speed Downlink Packet Access (standard)
HSUPA	High Speed Uplink Packet Access (standard)
IEEE	Institute of Electrical and Electronics Engineers
IETF	Internet Engineering Task Force
IP-OFDMA	Internet protocol—orthogonal frequency multiple access

IR	infrared
IrDA	Infrared Data Association
ISM	industrial, scientific, and medical
ISO	International Organization for Standardization
ITU	International Telecommunication Union
JCA	Java connection architecture
kbps	kilobits per second
LAN	local area network
LDC	local distribution companies (gas utilities)
LF	low frequency
LMDS	local multipoint distribution service
LMP	link manager protocol
LNG	liquefied natural gas
LTE	Long Term Evolution (standard)
L2CAP	logical link control and adaptation layer protocol
MAC	media access control
Mbps	megabits per second
MGW	media gateway
MHz	megahertz
MILS	military standard
MIP	mobile IP
MMDS	multichannel multipoint distribution service
MSC	mobile switching center
MSS	mobile subscriber station
NFC	near field communication (technology) or Near Field Communication (initiative)
NGLs	natural gas liquids
NLoS	non–line of sight
OEM	original equipment manufacturers
OFDM	orthogonal frequency division multiplexing
ONS	Object Name Service
PASC	Portable Application Standards Committee
PCMCIA	Personal Computer Memory Card International Association
PDA	personal digital assistant
PHY	physical (as in three physical layers)
POS	point of sale
psi	pounds per square inch

psig	pounds per square inch gauge
QoS	quality of service
RAN	radio access network
RF	radio frequency
RNC	radio network controller
ROI	return on investment
ROIC	return on invested capital
SCADA	supervisory control and data acquisition
SCO	synchronous connection oriented
SGSN	serving GPRS support node
SIG	Bluetooth Special Interest Group
SKU	stock keeping units
SNMP	simple network management protocol
SS	service set
SSID	service set identifier
STAC	slotted terminal adaptive collection
STAs	WLAN client stations
TDMA	time division multiple access
TG	task group
3G	third generation (mobile networks)
UE	user equipment
UHF	ultra-high frequency
UID	unique identifier
ULA	upper layer application
UMA	unlicensed mobile access
UMTS	Universal Mobile Telecommunications Standard
UNII	unlicensed national information infrastructure
URLs	Internet uniform resource locators
USAF	U.S. Air Force
USB	Universal Serial Bus
USB-IF	Universal Serial Bus Implementers Forum
UTRAN	UMTS terrestrial radio access network
UWB	ultra-wideband
VLR	visitor location register
WAP	wireless application protocol
WDS	wireless distribution system
Wi-Fi	wireless fidelity
WiMAX	Worldwide Interoperability for Microwave Access

WLAN	wireless local area network
WMAN	wireless metropolitan area network
WPAN	wireless personal area network
WWAN	wireless wide area network

Index

A

B

C

D

E

F

G

H

I

K

L

M

N

O

P

Q

R

S

T

U

V

W

X

Y

Z

About the Authors

Dipankar Sen has 25 years of experience in the water and energy markets. He is actively involved in design, operations, and strategic planning at utilities. Additionally, he focuses on technology development and adoption. He has advised several early stage companies, including Entex (www.entexinc.com) and Suganit (www.suganit.com), and developed products and processes for alternative energy and water reuse. He has created open-source tools to help a global user base with development and optimization of membrane and biofilm technologies (www.aquifas.com). He has an MBA from the Wharton School, a PhD from Virginia Tech, where he was a Via Research Fellow, and is a graduate of the Indian Institute of Technology, Kharagpur.

Prosenjit Sen is an entrepreneur with more than 20 years of experience with high tech and Internet companies in management, strategy, and technology. He is currently the CEO of Print2Click, a company based in Silicon Valley that develops software to drive transactions from media content and advertisements using mobile technology. He was part of the founding team of Informatica Corporation (NASDAQ: INFA), the leader in data integration software. He has held senior management and technical positions at SaleMinder, Unisys, Hewlett Packard, and Interval Research. He has a BS in electronics engineering from the Indian Institute of Technology, Kharagpur.

Anand M. Das is founder and CTO of Commerce Events (www.commercevents.com). He is one of the leading practitioners in RFID, helping Commerce Events develop solutions for the DOD, retail and utilities. He is a founding member of EPCglobal and the co-author of the book *RFID Security*. Previously, he was a vice president with SAIC where he led the RFID practice, a senior executive at BEA Systems, and held technology positions at Autodesk and Intergraph. He has an MS from Columbia University and obtained his undergraduate degree at Indian Institute of Technology, Kharagpur. He has served as a past chairman of Northern Virginia Technology Council (www.nvtc.org) and is an active member of AFCEA (www.afcea.org).

Additional Contributors

Ajay Bakre is a research scientist at Intel in Bangalore, India. He completed his PhD at Rutgers University, New Jersey, in 1996. He specializes in IP/ATM networking, broadband wireless access and storage networking. He is a member of Project Review and Steering Group, evaluating the implementation of a "Nationwide QoS Network Testbed" funded by Ministry of Communications and IT, Govt. of India. His current interests include wireless access technologies such as WLAN and WiMAX, Mobile IP, and multimedia networking applications.

Gary Panebianco is a vice president of natural gas operations and construction with National Grid. He has been a key figure in the natural gas industry for 30 years, managing operations and delivery of natural gas over a 4,500 square mile area with more than half a million customers. Gary obtained his undergraduate in engineering and MBA from Syracuse University, New York.

S.C. Sen is an ex-director, technical, with the National Hydro Electric Power Corporation, India. He has more than 45 years of experience in the electric industry and has played a leadership role in shaping the rapidly growing electric power industry in South Asia.